T0212858

Environmental Governance in China

Governance and Public Policy in China

Volumes published in this Brill Research Perspectives title are listed at *brill.com/rpgp*

Environmental Governance in China

State, Society, and Market

By

Jesse Turiel
Iza Ding
John Chung-En Liu

BRILL

LEIDEN | BOSTON

Library of Congress Control Number: 2017954630

Typeface for the Latin, Greek, and Cyrillic scripts: "Brill". See and download: brill.com/brill-typeface.

ISBN 978-90-04-35991-8 (paperback)
ISBN 978-90-04-35992-5 (e-book)

Originally published as Volume 1(2) 2016, in *Governance and Public Policy in China*, DOI 10.1163/24519227-12340002.

Contents

Author Biographies

Jesse Turiel is a Ph.D. student in the Earth and Environment Department at Boston University. He received dual Bachelor's degrees in Biology and Geography from Syracuse University, and his current research focuses on the interactions between public opinion and environmental governance in China. Along with Professors Anthony Saich and Edward Cunningham, he is currently administering a multi-annual public opinion survey of Chinese citizens with funding from the Harvard Ash Center for Democratic Governance. Turiel has conducted research on a broad range of energy and environmental issues, and has experience teaching courses in sustainable development, renewable energy, and international economics.

Iza Ding is an Assistant Professor of Political Science at the University of Pittsburgh. Her research examines post-Socialist political and economic development, with a substantive focus on bureaucracy, public opinion and environmentalism, and a regional focus on Asia and Central and Eastern Europe. She received her Ph.D. in Government from Harvard University in 2016, and B.A. in Political Science and Russian and Eastern European Studies at the University of Michigan in Ann Arbor in 2009.

John Chung-En Liu is an Assistant Professor of Sociology at Occidental College. During 2015–2016, he was the China Energy Policy Postdoctoral Fellow at the Ash Center for Democratic Governance and Innovation at Harvard Kennedy School. Liu received his Ph.D. in Sociology from the University of Wisconsin-Madison, joint Master's degrees in Economics and Environmental Management from Yale University, and a Bachelor's degree in Chemical Engineering from National Taiwan University. His research draws from economic and environmental sociology to study climate change governance. He has also published on the environmental public opinions in China.

Environmental Governance in China: State, Society, and Market

Jesse Turiel
Department of Earth and Environment, Boston University
jsturiel@bu.edu

Iza Ding
Department of Political Science, University of Pittsburgh
izading@gmail.com

John Chung-En Liu
Department of Sociology, Occidental College
chungenliu@oxy.edu

Abstract

This article provides an analytical overview of major works on the topic of environmental governance in China, with a particular emphasis on studies examining policies during the reform era (post-1978). We begin by exploring the rise of China's "environmental state" and the various institutional and political factors that shape state behavior. Next, we describe the complex relationship between the Chinese state and society, analyzing studies related to environmental public opinion, citizen action, nongovernmental organizations (NGOs), green civil society, the role of the media, and China's judiciary. Finally, we conclude by reviewing research on market-based mechanisms of environmental governance in China, including emissions trading schemes, environmental transparency, corporate information disclosure, and green finance.

Keywords

China – civil society – environment – governance – market – policy

1   Introduction

Since the beginning of the reform era (post-1978), the People's Republic of China (PRC) has enjoyed four decades of nearly uninterrupted economic growth, transforming a backward and isolated country into an emerging superpower with the world's second-largest economy. Although China has lifted hundreds of millions out of poverty, such an extended period of breakneck economic development has not come without a price. Rapid urbanization, industrialization, and growth in the gross domestic product (GDP) have ravaged China's natural environment and generated serious public health concerns. In 2014 alone, air pollution contributed to the death of 1.6 million Chinese citizens, accounting for nearly 30 percent of air pollution-related deaths worldwide (Brauer 2016; Rohde and Muller 2015). Of China's 560 million urban dwellers, less than 1 percent breathe air considered safe by European Union standards, while in the countryside, more than 300 million lack access to clean drinking water (Piovani 2015). Along China's frontier, the pace of deforestation and desertification continues to accelerate, and even the country's most pristine regions are showing signs of habitat destruction and biodiversity loss (Elvin 2008). On the global stage, China is now the world's largest emitter of greenhouse gases, releasing more carbon dioxide (CO_2) into the atmosphere than the United States, Russia, and India combined (Emissions Database for Global Atmospheric Research 2016).

The human cost of these environmental challenges is staggering. In numerous cities, towns, and villages across China, stories abound of rising rates of lung cancer, leukemia, asthma, and chronic respiratory infections (Zhao et al. 2010; Millman, Tang, and Perera 2008). In terms of economic damage, the costs of China's environmental degradation are less precise, though even conservative estimates peg the number between 3.5 percent and 8 percent of annual GDP (World Bank 2007). With each new "cancer village" or "airpocalypse" event covered by major domestic and international news agencies, the topic of environmental pollution in China has gained increasing traction among academics, government officials, business leaders, and the general public. Its significance as a source of public unrest and dissatisfaction is evinced by the growing number of environmental protests over the past decade (Hoffman and Sullivan 2015), as well as by the large majority of citizens who say that they would choose pollution control over economic development should the two priorities come into conflict (Ding 2016).

With so much attention now focused on the state of China's environment, the country's leaders must grapple with the Herculean task of responding to the social, economic, and political challenges caused by pollution. In order to better understand these challenges, we analyze China's system of environmental governance through the lens of state, society, and market. The overarching goal of this article is to identify the key debates, paradigms, and emerging research directions in the study of environmental governance in China. Given the breadth of the topic and our page limit, we do not seek to offer an exhaustive review of the literature, and, to our regret, several important works are necessarily left out. The article is divided into three main sections.

The first section analyzes state environmental protection behaviors. It begins with a brief overview of imperial and Maoist environmental legacies and then delves into research focused on the development of China's environmental state during the reform era. The second section provides a synopsis of nongovernmental environmental actors in China and the ways in which they interact with the Party-state. In examining the nature of environmental state-society relations, this section analyzes existing literature concerning the role of public opinion, environmental protests, environmental nongovernmental organizations (ENGOs), the media, and China's legal system. The third section offers a look at the business and market forces at play in shaping China's response to environmental challenges. It reviews the literature on market-based policies, such as cap and trade, green financing, and payment for ecosystem services. It also examines the vital role of information in determining environmental outcomes and details how both private and state-owned corporations respond to regulatory pressure from the government and social pressure from the general public. Finally, the concluding section provides a brief synopsis of the current state of China's environmental reforms and identifies some key areas for future environmental research.

2   China's Environmental State

Imperial and Maoist Legacies

Although this article focuses primarily on environmental governance during the reform era, it is nevertheless important to set the stage by dispelling some popular myths about environmental change and governance in China. One myth is that environmental degradation is strictly a reform-era phenomenon. This notion could not be further from the truth (Edmonds 1994; Elvin 1998, 2008; Marks 2011; Qu and Li 1994; Smil 1984). Elvin (1998) notes that, in imperial China, population expansion, agricultural development, and infrastructure projects made the Chinese environment "one of the most transformed in the pre-modern world" (p. 742). The political consequences of environmental change are also well documented. For example, Wakeman (1985) connects the brutal toppling of the Ming dynasty (1368–1644) by Manchus to pressure on the northern tribesmen to locate new food sources during the Little Ice Age: a period between the sixteenth and nineteenth century when average global temperatures cooled substantially.[1] In addition, scholars at Beijing Normal University found a statistical correlation among low temperature, subsequent famine, and peasant rebellions throughout imperial China (Fang et al. 2015). By late-imperial times, the depletion of natural resources as a result of population expansion and agricultural development had evoked widespread concern among the literati (Elvin 1998). By then China had lost substantial swaths of its forests, and natural disasters were frequent (Elvin 1998).

In the search for explanations of the human causes of environmental degradation, cultural explanations (East and West) have borne the blame. Carolyn Merchant (1980) argues that the scientific revolution and market-oriented culture of the West marked the "death of nature" through the objectification, dissection, and domination of the natural environment. Similarly, scholars of China's environment have expressed the view that traditional Chinese culture, as exemplified by Confucianism, pits men against nature (Economy 2004; Shapiro 2001). This notion is articulated by the popular idiom stating that "man must triumph over nature" (*ren ding sheng tian* 人定胜天). However, as Economy (2004) notes, evidence of environmental governance can be traced back to at least the Western Zhou dynasty (1050 BCE–771 BCE), and the *Book of Rites* (*Zhou li* 周礼) documents that local officials were responsible for protecting their local rivers, mountains, forests, and animals (p. 29). Confucian teachings also call for the reverence of nature, while other Chinese spiritual traditions, such as Buddhism and Daoism, advocate a harmonious relationship between people and nature (pp. 33–36).

In the Maoist era (1949–1976), environmental problems were overshadowed by immense human suffering, yet studies by Qu (1994) and Shapiro (2001) vividly demonstrate the grievous injuries inflicted on nature during this period. During the Great Leap Forward (1958–1962), massive forests were felled to fuel backyard steel furnaces in a blind attempt to surpass Great Britain in steel production. During the 1958 "Exterminate the Four Pests" campaign, Mao, who knew little about wildlife or ecology, called for the extermination of sparrows in the belief that, because sparrows eat crops, crop yields would naturally improve as a result of their extermination. However, the systematic extermination of hundreds of millions of sparrows led to the proliferation of locusts—an insect eaten by sparrows—and a significant drop in crop yields, exacerbating the

---

1 Although the Little Ice Age was not a manmade event, its worldwide political impact was tremendous.

Great Leap Famine (1959–1961). During the 1964 "On Agriculture, Learn from Dazhai" campaign, localities blindly conformed their natural environments to resemble that of Dazhai—a hilly, self-sufficient agricultural community in rural Shanxi Province. Writing about the campaign, Qu (1994) observes that "regardless of topography, grain production became an all-important priority.... Large forested areas were either destroyed to produce grain or neglected, aggravating hydrological cycles and soil erosion" (p. 61).

Maoist policies have left three major marks on the current state of environmental governance in China. First, the development-first mind-set remains highly influential, among both government officials and the general public, in shaping attitudes about environmental protection. Only recently has environmental protection moved to the forefront of China's policy agenda. Second, policy decentralization and the *nomenklatura* system continues to encourage fierce competition among local officials (Kung and Chen 2011), which has contributed to rapid economic growth (Zhou 2007). Third, Maoist campaign-style governance is still prevalent in today's environmental management practices (Liu et al. 2014; van Rooij 2006). These massive, ad-hoc government efforts—ranging from annual tree-planting campaigns to crisis management responses—are reminiscent of Maoist campaigns and are sometimes wasteful and ineffective (Economy 2014).

The Ideological Genesis of Environmental Governance

China's modern environmental state was born in the 1970s. In 1972, the United Nations Conference on the Human Environment convened in Stockholm, with China attending as a new UN member. The meeting's report urged China to pay more attention to its domestic environmental problems, and in August 1973, China's State Council (the Cabinet) convened the country's first national conference on environmental protection and passed "Some Decisions on the Protection and Improvement of the Environment"—the first policy document to regulate environmental protection in the People's Republic of China.[2] The Decision established several of China's key environmental governance principles, including "balancing economic development and environmental protection," "preventing pollution first; combining prevention and management," and "whoever generates pollution manages it." It also set the stage for the passage in 1979 of China's first Environmental Protection Law.

---

2   "Huanjing baohu zhuangkuang 环境保护状况 [Situation of Environmental Protection]," Government of the PRC, July 27, 2005, accessed June 30, 2016, http://www.gov.cn/test/2005–07/27/content_17757.htm.

Backlash against environmental protection was strong from the outset, especially among managers of state-owned enterprises who were under pressure to fulfill production targets. Meanwhile, public awareness about environmental protection remained low because the notion of environmental protection was novel to citizens of the PRC.[3] To raise awareness, state media gradually increased their use of the phrase "environmental protection" (*huanjing baohu* 环境保护).[4] At the same time, new environmental policies were promoted through a series of "patriotic hygiene campaigns," publicizing the link between pollution and public health. Furthermore, the state framed pollution control as a demand by the public (Ding 2016, pp. 35–36). After the first national conference on the environment, the *People's Daily* made a point of publishing a letter from two teachers in Hubei Province that criticized the neglect of the "three wastes" (wastewater, waste air, and solid waste) by factories and called for the government to "value the principle of being responsible to the people and attach high importance to pollution control" (Tao and Ai 1973).

As early as the 1970s, the central government recognized the pitfalls of the developmental trajectory followed by countries that had previously industrialized. The 1973 Decision's emphasis on ex ante pollution prevention, rather than ex post abatement represented a sea change in China's approach to environmental governance and demonstrated the central leadership's awareness of the problems associated with the "pollute first, control later" (*xian wuran, hou zhili* 先污染, 后治理) model.

Development of China's Environmental State

During the first decade of economic reform, China's modern environmental state emerged, albeit with some growing pains (Ross 1992, p. 631). In the same year as the official initiation of reform and opening up policies (1978–1979), the State Council enshrined environmental protection in the Chinese

---

3   During the first two decades of the PRC, criticizing environmental pollution was tantamount to smearing socialism—a system that could do no wrong. After the most radical stage of the Cultural Revolution, discussing the negative externalities of development was no longer taboo.

4   "Environmental protection" first appeared in *People's Daily* in 1972; see "Wo guo daibiaotuan tuanzhang Tang Ke zai lianheguo renlei huanjing huiyishang fayan chanshu wo guo dui weihu he gaishan renlei huanjing wenti de zhuzhang 我国代表团团长唐克在联合国人类环境会议上发言阐述我国对维护和改善人类环境问题的主张 [Tang Ke Spoke on Behalf of Chinese Delegate about Our Country's Stance on the Protection and Improvement of Human Environment at the United Nations Conference on the Environment]," *People's Daily*, June 11, 1972; and Xin Fang, "Jingji fazhan he huanjing baohu 经济发展和环境保护 [Economic Development and Environmental Protection]," *People's Daily*, June 16, 1973.

constitution, subsuming it under the Four Modernizations and declaring once again that China would not walk down the "pollute first, clean up later" path (State Council Leading Group on Environmental Protection 1978).

Two main features marked the development of China's environmental state. The first was the rapid promulgation of laws, policies, and regulations concerning environmental protection; and the second was the expansion of China's bureaucracy charged with environmental protection. As mentioned earlier, China's first national environmental protection law, formally titled the Environmental Protection Law of the People's Republic of China (Trial Implementation), was passed by the National People's Congress (NPC) in 1979.[5] Among its many provisions, the law allows local governments to pass their own laws, policies, and regulations on environmental protection. Subsequently, dozens of national laws and hundreds of local regulations were adopted over the following decades. As of 2016, China has 20 national environmental laws passed by the NPC, nearly 200 national environmental regulations issued by the State Council, and more than 1,000 local environmental regulations adopted by local people's congresses.

Coupled with this proliferation of laws and regulations is the impressive growth in the size, capacity, and status of China's environmental bureaucracy (Jahiel 1998, p. 768). Before the 1970s, the Ministry of Public Health was in charge of environmental policy. However, in May 1974, the State Council established the Environmental Protection Leading Group to coordinate environmental policy-making, ushering in a nascent environmental bureaucracy. The Leading Group consisted of officials from bureaucracies responsible for economic planning, industrial policy, agricultural policy, public health, traffic control, and water resource management. Many provinces followed suit by establishing their own environmental leading groups, often called the Office for the Control and Management of the Three Wastes.

The first environmental administrative bureaucracy—the Environmental Protection Bureau (EPB)—was established in 1982 and housed within the Ministry of Urban and Rural Construction (MURC). Two years later, the bureau was renamed the National Environmental Protection Agency (NEPA); then, in 1988, NEPA was separated from the MURC, and its bureaucratic rank was raised to a half-notch below ministerial level (Jahiel 1998, p. 769; Sinkule and Ortolano 1995).

The development of China's environmental state experienced a temporary setback in 1992 (Jahiel 1998). During a tour of southern China, Deng Xiaoping called for localities to speed up the pace of economic development. Thereafter

---

5 The Environmental Protection Law (EPL) was officially adopted in 1989 and revised in 2014.

followed a surge in the growth of township and village enterprises (TVES), whose pollution practices were largely unsupervised. Pollution control was sidelined in 1993 and 1994, and many local EPBS were demoted or abolished (Jahiel 1998, p. 773). However, this institutional setback did not last long. In 1998, NEPA was restyled as the State Environmental Protection Agency (SEPA) and promoted to the ministerial level, providing it with significantly enhanced bureaucratic authority. A decade later, SEPA was elevated to cabinet-level status and renamed the Ministry of Environmental Protection (MEP).

Between 1995 and 2004, employment in China's environmental bureaucracy nearly doubled (from 88,000 to more than 160,000), while state investment in environmental protection increased from 0.6 percent of GDP in 1989 to 1.4 percent in 2004 (Mol and Carter 2006). Accordingly, the policy jurisdiction of EPBS expanded from the management of the "three wastes" in the 1970s to their current plethora of responsibilities, including issues as wide-ranging as biodiversity, energy efficiency, and nuclear safety.

In order to carry out environmental policies and control local pollution, EPBS rely on three main procedural levers (also known as the "three magic weapons") (Sinkule and Ortolano 1995). These three levers, which are implemented at different points of the pollution process, include the "Three Simultaneities" (*santongshi* 三同时) mandate, the environmental impact assessment (EIA) system, and pollution discharge fees (Sinkule and Ortolano 1995). The Three Simultaneities mandate, first introduced in the 1973 Decision, specifies that the design, construction, and operation of pollution control mechanisms must occur concomitantly with the design, construction, and operation of the parent industrial project. New major construction projects and enterprise operations must be filtered through the EIA system, allowing the environmental impacts of development projects to be assessed. Individual enterprises must then use EIA reports to seek approval for new construction from their local EPBS. If approved, the EPB gives the enterprise a pollution permit, specifying the quantity of each type of pollutant that the enterprise can legally discharge. Finally, the enterprise must agree to pay a fee for the emission of pollutants into the environment, and discharging more than the allotted amount can result in fines, sanctions, or even forced closures.

Environmental Policy-Making in the Reform Era

Since the 1980s, a rich academic literature on environmental governance in China has emerged alongside increasing general concern about environmental degradation. State actors, due to their dominance in the political system, were the first to attract scholarly interest. Some of the central questions in this literature include: What factors shape the environmental policy-making process at the central and local levels, and what determines how policies are implemented? Also, how does the state deal with increasing pressure for environmental protection, and what kind of institutional and extra-institutional forces affect state behavior? To answer these questions, scholars have explored a variety of factors relating to state behavior, such as the structural features of China's political system, incentive structures for local officials, bureaucratic politics, and the role of nongovernmental actors.

Lieberthal and Oksenberg's (1998) seminal work explains policy-making in China using a model of "fragmented authoritarianism." Looking at the structural factors that shape energy policy, they write that "the fragmented, segmented, and stratified structure of the state promotes a system of negotiations, bargaining, and the seeking of consensus among affected bureaucracies" (p. 3). In essence, their argument contains three main points: first, decision-making authority in the Chinese political system is fragmented and disjointed; second, multiple bureaucracies participate in setting the policy agenda; and, third, bureaucracies bargain and compete to shape policy outcomes (Lieberthal and Lampton 1992). Although articulated more than twenty years ago, fragmented authoritarianism endures as a quintessential framework used by many scholars when analyzing policy-making in China.

Unsurprisingly, bureaucratic politics is at the heart of environmental policy-making. In attempting to defend and enlarge their decision-making authority, different government agencies frequently engage in turf wars for administrative power over specific policy domains. Jahiel (1998) notes that, since the 1990s, international environmental conferences and funding for environmental protection efforts have motivated bureaucracies to "[scramble] for a piece of the environmental pie" (p. 786). Bureaucracies such as the MEP, the National Development and Reform Council (NDRC), the Ministry of Commerce, and the Ministry of Science and Technology have all competed for control over various environmental policies, such as clean production, energy conservation, and regulations concerning genetically modified organisms (GMOs) (Mol and Carter 2006). In this fight, powerful economic agencies such as the NDRC often emerge victorious, as evidenced by China's recent attempts to address climate change.

In addition to powerful bureaucracies, state-owned enterprises (SOEs) also have substantial influence in the environmental policy-making process. Indeed, much of China's energy sector is still dominated by a handful of large SOEs. For example, the China National Petroleum Company (CNPC) and the China Petroleum and Chemical Corporation (Sinopec) together account for 72 percent of the country's annual crude oil output, 75 percent of annual refined oil output, and nearly 90 percent of annual natural gas extraction (Wang

2016). China's power generation sector is more diverse; since 2003 the "big five" SOEs account for approximately 50 percent of annual electricity generation (Cunningham 2015). The transmission and distribution of energy, however, remain a virtual monopoly, with roughly 80 percent of annual electricity delivered by the State Grid Corporation of China (SGCC), and the rest controlled by China Southern Grid. Thus, throughout the 1990s and 2000s, large SOEs were often able to use their outsize political influence to reshape energy regulations or simply ignore them (Downs and Meidan 2011).

In China, political linkage between businesses and government leaders remains strong, and, since the early days of the PRC, scholars have noted the presence of factions within the elite leadership circles of the Chinese Communist Party (CCP). One of these factions was the "Petroleum Gang," established under the Minister of Petroleum Yu Qiuli in the late 1950s (Page et al. 2013). Under the administration of President Jiang Zemin (1993–2003), the power of this group was revived under Zeng Qinghong (a Politburo Standing Committee member and also a key associate of Jiang's "Shanghai Clique") and eventually transferred to Zhou Yongkang. As head of the CNPC between 1996 and 1998, Zhou was able to exert strong influence over policies and appointments in the energy industry, and this influence continued even after Zhou left the industry and ascended to the Politburo Standing Committee in 2007 (Liao 2014). Meanwhile, other CCP leaders also carved out networks of influence in the energy sector, most notably the technocratic Premier Li Peng (Peng, among other things, served as a Party secretary of the Beijing Electric Power Administration, oversaw the construction of the Three Gorges Dam in 1994, and later helped his children become the directors of two of the "big five" electricity producers) (Barboza and LaFraniere, 2012). These iron-clad ties between industry and government officials have made it especially difficult to enact meaningful environmental reforms, and some observers have argued that breaking up political factions is the only way to move forward with environmental protection.

In recent years, engagement in the policy-making process has become possible for extra-institutional participants, and an emerging literature points to the fact that the Chinese state selectively responds to societal input (e.g., Nathan 2003; Reilly 2013; Stockmann 2012; Weller 2008). Wang (2008) observes that environmental policy-making in China is often initiated by citizens outside the government, either through individual lobbying or advocacy groups. Mertha's research on hydropower policy-making (2008, 2009)—building on the fragmented authoritarianism model—also calls attention to the pluralization of the policy-making process, which has increasing involvement by unconventional actors, such as the media, activists, scientists, NGOs, and peripheral officials. Van Rooij et al. (2016) also show that, although the state grants new actors access to environmental regulatory agencies, it simultaneously keeps these actors under its tight control.[6]

Finally, the international diffusion of environmental institutions has played an instrumental role in shaping environmental policies in China. In fact, in many instances, the state has actively sought to learn from the experiences of foreign countries when designing China's environmental laws and policies. For example, China's national water quality standards and EIA system were drawn from American blueprints, while its system of pollution discharge fees was borrowed from similar programs in France and Germany (Sinkule and Ortolano 1995). Moreover, international NGOs, such as the Environmental Defense Fund, Resources for the Future, and the China Council for International Cooperation on Environment and Development have all been active in advising the Chinese government on the development of programs for emissions trading (*paiwuquan jiaoyi* 排污权交易), carbon trading (*tanpaifang jiaoyi* 碳排放交易), and the calculation of a Green GDP.

Challenges in Policy Implementation

Although Chinese environmental protection statutes and regulations still suffer from many flaws, most scholars acknowledge that China has made impressive strides in developing a strong and comprehensive set of environmental institutions (Carter and Mol 2013; Wang 2006). However, although these laws may appear strong on paper, problems associated with policy implementation abound. As Lieberthal and Oksenberg (1988) note, "much of the environmental energy generated at the national level dissipates as it diffuses through the multi-layered state structure, producing outcomes that have little concrete effect" (p. 3). In essence, China's decentralized political structure has permitted localities a great deal of flexibility in responding to central mandates, and this flexibility has often resulted in weak enforcement of environmental protection policies. According to Qi (2008), the problem of weak enforcement can be explained by a multitude of factors, including, but not limited to, the overall developmental strategy, regulatory loopholes, lack of funding, and administrative weakness in the environmental bureaucracy. However, three structural and institutional explanations stand out as the most fundamental: the incentive structure of local leaders, the fragmented nature of the governing system, and the top-down command-and-control approach to environmental governance.

At the local level, the decision to reduce emissions or strengthen enforcement is ultimately in the "first hands" (*yibashou* 一把手)—usually local Party

---

6 This literature on extra-institutional participants in China's environmental governance system is further explored below in the section "Environmental State-Society Relations."

secretaries. Existing research shows that the CCP controls the behavior of local cadres through the cadre evaluation system (*ganbu kaohe zhi* 干部考核制) (Chen et al. 2005; Huang 1985; Manion 1985; Whiting 2004).[7] Every year, the Party sets ranked targets—some hard, some soft—in key policy areas. Achievement (or underachievement) of these targets factors into cadres' annual performance reviews, which in turn are used in promotion decisions. As a result, performance targets are a powerful tool for guiding local officials in setting policy priorities throughout their tenure.

As early as the 1980s, environmental governance became one of the categories used to evaluate the performance of local officials (Ross 1992). The eleventh FYP (2006–2010) introduced specific objectives for pollution control, and the twelfth FYP (2011–2015) added targets for carbon emissions reduction and energy efficiency. However, Chinese elites hold a broad consensus that economic development is the country's top priority, a notion reinforced by the findings of various researchers. For example, Wang (2013) argues that environmental goals in the cadre evaluation system have only been prioritized to the extent that they benefit economic development and social stability. Similarly, Lo and Tang (2006) find that little importance is attached to environmental governance issues when the performance of local leaders is assessed. Even more strikingly, Wu et al. (2013) find that local government spending on environmental protection is uncorrelated with promotion odds for local officials, while spending on transportation infrastructure (which directly contributes to raising land values and GDP) increases the likelihood of promotion.

Another factor that affects the incentive structure of local leaders is their individual time horizons with respect to policy-making and implementation. For instance, a term-limited mayor or Party secretary is likely to set different policy priorities than a leader with no such time horizon. At the local level, government officials can remain in their post for only up to two terms (ten years), after which—with or without a promotion—they must move on to a new position, usually to a different locality. In practice, only a small percentage of local cadres finish two terms, and many are transferred to a different post before finishing their first term. Scholars argue that this constant rotation of officials into different posts allows the CCP to maintain oversight and control over local officials (Edin 2003; Huang 1996).

Frequent cadre turnover also disincentivizes local leaders from prioritizing meaningful environmental protection measures. Because the positive results of environmental protection typically are not immediately apparent, cadres with short tenures see little benefit in prioritizing environmental projects.

---

7   Others have challenged this view (e.g., Shih et al. 2012).

Eaton and Kostka (2014) find that, even when they do invest in pollution control, local leaders facing short time horizons prefer to implement costly show projects that generate short-term results while ignoring longer-term, sustainable alternatives. Furthermore, because most local officials do not expect to remain long at their current posts, they do not hesitate to take on debt to pay for costly development projects, because any budget deficits or negative environmental impacts can simply be left for their political successors. Thus most cadres end up paying more attention to the visibility and political benefits of local projects than their concrete environmental benefits (Eaton and Kostka 2014).

Furthermore, due to the interjurisdictional nature of many types of pollution and the lack of interjurisdictional collaboration in pollution control, environmental goals are often viewed as especially difficult, if not impossible, for local cadres to attain (Jahiel 1998). Environmental protection initiatives also tend to generate fewer opportunities for "gray" personal income for local leaders than more economically focused policies, such as infrastructure development. Thus environmental initiatives typically languish on the backburner of local government priorities. The one exception to this lack of emphasis occurs when environmental protection becomes an issue of social stability and local grievances lead to large-scale social protests. In this case, the "one-vote veto" (*yipiao foujue* 一票否决) system is applied to local cadres' performance reviews and the likelihood of promotion drops.

In addition to local first hands, directors of local EPBs form another relevant group of state officials (van Rooij 2003). Kostka's (2013) analysis of a biographical database of provincial EPB heads shows that they are highly embedded in local political networks and that this embeddedness improves EPB enforcement capacity while also diluting enforcement incentives. Kostka also finds that only one-fourth of provincial EPB heads are promoted through the ranks in the environmental bureaucracy. Instead, a clear majority originate from the staff of other industrial bureaus, the Communist Youth League, and SOEs, whose officials tend to be more development-oriented than career environmentalists.

When it comes to the implementation of environmental policy, local EPBs are the most important players, and a substantial amount of literature focuses on their operation (e.g., Ding 2017; Lo and Tang 2006; Ma and Ortolano 2000; Sinkule and Ortolano 1995). As mentioned earlier, China's environmental protection apparatus is marked by a clear fragmentation of authority (Jahiel 1998). Two types of fragmentation are key. The first is the overlap of political authority arising from the *tiao* (vertical)–*kuai* (horizontal) system, and the second is the fragmented distribution of environmental responsibilities across different bureaucracies. The *tiao-kuai* system channels authority through vertical, functional lines (from upper-level bureaucratic agencies to offices one level below them) as well as through horizontal, territorial lines (from local governments to local bureaucratic agencies). Because of this arrangement, most bureaucratic offices are subject to dual leadership. For example, a city-level EPB answers directly to both the city government and the provincial EPB. This *tiao-kuai* system creates a situation in which environmental agencies must serve two masters, often with different policy goals, resulting in policy implementation gridlock when vertical and horizontal lines of authority conflict with each other. In practice, because local governments appoint EPB officials and fund their operations, the horizontal ties between local bureaucracies and local governments are often stronger than the vertical ties between different layers of the environmental bureaucracy (Jahiel 1998; Lieberthal and Oksenberg 1988; Ma and Ortolano 2000). This typically results in growth-oriented policies, since local governments often rely on businesses—many of which are highly polluting—for tax revenue and employment. This leaves local EPBs in a weak bargaining position with respect to creating and enforcing environmental protection measures.

Authority over local environmental issues is also divided among various bureaucracies. For example, in a city, the police and the EPB are collectively responsible for controlling automobile exhaust fumes; the Bureau of Forestry and Water Resources is responsible for managing forests and surface water quality; the Bureau of Land Resource Administration is responsible for groundwater safety; the EPB is responsible for managing specific pollutants, such as chemical oxygen demand (COD), sulfur dioxide (SO_2), and nitrous oxides (NOX); and the NDRC is responsible for managing carbon emissions. According to Jahiel (1998), such fragmented authority can lead to the neglect of environmental issues by industrial bureaus, a lack of coordination among government agencies, and excessive competition among bureaus as they fight for environmental funds. Overall, incentives designed by central authorities for local officials have had a perverse impact on policy enforcement at the local level. As Ran (2013) writes, weak local implementation of environmental regulations is a direct result of the "central government's failure to encourage—politically, financially, as well as morally—local government officials to appropriately implement environmental policies."

Finally, central authorities have been using a command-and-control approach to manage the implementation of local environmental policy. Based on goals outlined in the FYP, the MEP sets nationwide pollution reduction targets for each pollutant at the beginning of each year. These targets are then passed down to provincial governments, who in turn distribute them to the cities and counties in the province. At the beginning of the year, local EPB directors sign a "target responsibility document" (*mubiao zeren shu* 目标责任制) with the local government, which outlines the specific goals that the EPB should achieve in the upcoming year. This command-and-control approach is often used by authoritarian regimes, and proponents of "environmental authoritarianism" argue that autocrats have an advantage vis-à-vis democratic regimes in managing environmental issues, due in large part to their greater insulation from interest groups and public opinion (Beeson 2010). However, Kostka (2016) warns of the difficulties associated with this approach, including local noncompliance, data manipulation, and the rigidity of targets that often fail to reflect local conditions.

Governance Style and Bureaucratic Strategies

A persistent characteristic of Chinese EPBs is their being poorly funded, understaffed, and administratively weak. Given these challenges, academics and policy-makers alike have attempted to understand how EPBs cope with mounting responsibilities and limited institutional support. Multiple strategies used by EPBs have been identified in the literature.

First, EPBS rely on economic incentives. During the 1980s and 1990s, local EPBs faced a shortage of funds, and many relied on pollution discharge fees to support the expanding scope of their activities (Jahiel 1998, p. 775). Lo and Tang's (1994) research on the management of water pollution in Guangzhou shows that making pollution fees and fines a part of the operating budget for the municipal EPB incentivized the agency to make "substantial efforts in enforcing water pollution regulations" and "produce[d] positive results in controlling industrial water pollution, recalling the water quality in Pearl River" (p. 59). Similarly, Ma and Ortolano's (2000) study of six municipal EPBs finds that they were highly motivated to implement pollution fees because they yielded substantial government revenues. In 1994, city EPBs received only 30–40 percent of their operating expenses from city governments, and some county and district EPBs received only 4 percent of their budget from county or district governments, with the rest coming from fees and fines (p. 124). However, this reliance on pollution fees can also create incentives for unnecessary levies designed to bring in revenue, rather than to control pollution (Jahiel 1998; Ma and Ortolano 2000, p. 125). It also frequently led to "backroom" arrangements that benefited both EPBs and local enterprises. For example, some county EPB employees became shareholders in local enterprises, and some enterprises recruited informants from local EPBs to obtain information about upcoming inspections. During the late 1990s and early 2000s, the State Council called for the establishment of "two separate lines for revenues and expense" (*shouzhi liangtiaoxian* 收支两条线), with pollution fees and fines being paid directly to local governments, instead of EPBs. The "separate lines" reform made local EPBs less reliant on pollution fees and fines as a means of income, and research conducted in Guangdong by Lo and Tang (2006) suggests that such efforts have been largely successful.

Second, EPB officials—to the extent that they have an incentive to advance their agency's interests—engage in lobbying and coalition building (Zhou et al. 2013). For example, Ross (1992) observes that SEPA's political status improved during the late 1980s because it obtained strong backing from the powerful State Science and Technology Commission (SSTC), as well as support from key members of the State Council (p. 630). Likewise, Jahiel (1998) notes that EPBs may undertake educational campaigns to convince other bureaus of the importance of their work. They may also court local political allies or attempt to gain support from the judicial system (p. 766). For example, in 1995, NEPA forged an alliance with the People's Bank of China (PBOC), which subsequently agreed to refuse credit to firms with poor records of environmental performance (p. 778). Ma and Ortolano (2000) also observe that local EPBs use "cooperative approaches" to enhance their influence, often inviting enterprise representatives and industry officials to meetings for "a sincere discussion" in order to hammer out solutions to compliance problems (p. 121). Finally, some EPBs run training workshops and pay "informal visits" to factories, ostensibly to "discuss pollution problems and provide assistance" (p. 122).

Third, faced with limited resources, local EPBs use various short-term coping strategies to meet their bureaucratic goals. Many scholars have attempted to conceptualize these coping strategies, examples of which include "campaign-style governance," "crisis management," "muddling through," and "performative governance." Campaign-style governance is a key feature of reform-era environmental management. As mentioned earlier, it is a relic of Maoist politics (Heilmann and Perry 2011) and can be observed in many current policy domains, such as "constructing the new socialist countryside," environmental protection, patriotic education, and efforts to combat corruption. Although most scholars agree that campaign-style governance suffers from significant drawbacks (Economy 2014), other researchers argue that these efforts can sometimes be effective. For example, Zhao and Ortolano (2010) report that industrial air pollution was reduced during the eleventh FYP as a result of widespread emissions reduction campaigns. Similarly, Liu et al. (2015) claim that environmental campaigns can have a lasting impact on those involved and may accelerate future changes.

Using a slightly different conceptual framework, Economy (2013) describes the style of China's environmental governance as one of "crisis management." Crisis management involves quick, decisive, and temporary measures to mitigate the most urgent symptoms of problems that may be socially or politically destabilizing. Examples of crisis management strategies include extravagant tree planting campaigns, the construction of massive ecocities, and acceding to citizen demands when protests occur (p. 185). Economy contends that these crisis management techniques are often wasteful and ineffective. For example, she notes that China's highly publicized tree-planting campaigns have an estimated failure rate of 85 percent (p. 188) and that half the central funds designated for environmental protection end up diverted to local infrastructure development projects (p. 189). Although some crisis management techniques may provide temporary symptomatic relief (e.g., shutting down factories near cities where international conferences are being held), they do not typically generate incentives for long-term, meaningful changes at the local level.

Zhou et al. (2013) use the term "muddling through" to describe the behavior of local EPBs. The authors define muddling through as "a reactive response to multiple pressures, constant readjustments and a focus on short-term gains" (p. 120). Because of the fragmented nature of China's environmental governance system, local EPB officials receive "time-varying, multiple, and sometimes conflicting pressures imposed from the top" (p.124) and are forced to rely on incremental approaches to cope with the unforeseeable consequences of their actions.

Finally, Ding (2016) observes that, although local EPBs are sensitive to directives imposed by higher authorities, they are equally sensitive to the pressure of local public opinion. Using ethnographic evidence, Ding demonstrates that local EPBs engage in a process called "performative governance"—the state's theatrical deployment of language, symbols, and gestures to signal the provision of "good governance" to its citizens. Examples of performative governance include responding to citizen complaints in a timely fashion, demonstrating sincere concern in the language used by public officials, patiently explaining relevant policies, and exhibiting careful and in-depth investigation into prominent issues. When local bureaucracies lack the political resources to meet citizen demands in the face of high pressure from public opinion, they often engage in performative governance, despite knowing from the start that citizen concerns will not be fully addressed (Ding 2017).

Institutional Innovations for Pollution Control

In recent years, central authorities have launched multiple institutional innovations—including the Urban Environmental Quality Examination System, the National Environmental Model City competition, the Open Environmental Information (OEI) regulations, and the short-lived Green GDP program—in order to stimulate local environmental initiatives (Economy 2006). The central government has also increased funding for local environmental projects (Mol and Carter 2006). These innovations, however, do not always achieve their objectives. For example, many localities have launched massive clean-up campaigns to obtain designation as National Environmental Model Cities, only to reduce pollution control efforts after the status is achieved. Furthermore, during the mid-2000s, the central government's ill-fated Green GDP program— developed with the National Bureau of Statistics and SEPA—was ultimately crippled by strong local resistance, with some localities outright refusing to provide vital environmental statistics. Many towns and counties have also developed an unhealthy reliance on environmental funds from the central government, and have even purposely kept their rivers dirty in order to continue receiving financial assistance (Ding 2016).

In recent years, local EPBs have been permitted a greater degree of autonomy from local economic interests and more freedom to fine-tune methods of environmental governance to better suit local conditions (Lo and Tang 2006). Greater decentralization in environmental governance has led to a plethora of local innovations. For example, the "river chief" system (*hezhang-zhi* 河长制) in Zhejiang and Jiangsu Provinces holds local leaders personally responsible for maintaining specific segments of local rivers. Reforms that seek to break up the strong political linkages between SOEs and local governments can also serve to grant EPBs more administrative autonomy (Mol and Carter 2006).

During the 2013 meetings of the NPC and the National People's Political Consultative Conference, the State Council unveiled plans to reshuffle various ministries and streamline their responsibilities. In the environmental realm, the state planned to consolidate the duties of environmental protection into a single bureaucracy (the MEP), while also expanding its administrative capacity. This, along with an increased emphasis on vertical management (*chuizhi guanli* 垂直管理), signaled that central leaders were intent on making local EPBs less reliant on local governments and more responsive to direct supervision from provincial EPBs.

In September 2016, the power to nominate city-level EPB directors was transferred from city governments to provincial EPBs, although city "approval" was still required to confirm nominees. If all goes according to plan, this new rule will make local EPB directors significantly more accountable to vertical supervision. Furthermore, city-level environmental monitoring stations will soon be absorbed by provincial EPBs, meaning that provincial EPBs will assume responsibility for the funding and personnel decisions of local monitoring stations (Xinhuanet 2016). This policy shift was designed to reduce the ability of local leaders to manipulate environmental data and influence the action of local EPBS.

In recent years, the government has also worked to expand the role of citizen participation in the governing process. As environmental complaints grow more frequent, local EPBS are becoming increasingly responsive to them, and some EPBS have even held public hearings to mediate conflicts between citizens and local businesses. Citizens are also encouraged to participate in the review of environmental impact assessments, although only a small percentage of projects require compulsory citizen participation (Economy 2006). Although environmental concern and participation are undoubtedly on the rise, scholars disagree about the exact nature of environmental state-society relations. The following section provides an overview of existing debates in the literature on opportunities for, and limitations to, citizen engagement in the China's environmental governance process.

3   Environmental State-Society Relations

Searching for an Emerging Environmental Movement in China

As the Chinese government continues to debate the scale and scope of future reforms, it must contend with a nationwide upsurge in popular concern over environmental issues, not just among government officials and the media but also among China's nearly 1.4 billion citizens. Between 1996 and 2011, the number of "environmental mass incidents" in China increased at an average annual rate of 29 percent, capped by a stunning increase of 120 percent in 2011 alone (at which point the government stopped reporting official statistics) (Balme and Tang 2014). Furthermore, the Chinese Academy of Social Sciences estimates that, between 2000 and 2013, half the large-scale protests were triggered by concerns over environmental pollution (Lin and He 2014). Indeed, several of these environmental demonstrations, including the 2004 Nu River protests and the 2007 Xiamen anti-PX plant protests, have been singled out by researchers as evidence of society's ability to shape official policy in China (Jian and Chan 2016).

In addition to the growth of popular demonstrations, scholars have also noted the rapid proliferation of ENGOs, of which more than 7,000 are currently registered with the Ministry of Civil Affairs (Zhang, Mol, and He 2016). These ENGOs vary greatly in their level of independence from and cooperation with government authorities, but virtually all favor an increased role for the general populace in environmental decision-making. Even the central government appears to have acknowledged the need for greater openness and transparency in environmental affairs, passing the Environmental Information Disclosure Measures (EIDMS) in 2008, which require government authorities and serious industrial polluters at all levels to disclose environmental information to the public (Zhang et al. 2016). These measures were followed by a strengthening in 2012 of the country's National Ambient Air Quality Standards (NAAQS), and in 2014 by a revised National Environmental Protection Law, which, among other things, enhances the ability of local EPBS to monitor and punish polluters (Feng and Liao 2016; Zhang, He, and Mol 2015). However, even polluters that are ignored or overlooked by government authorities are not immune from the court of public opinion, as both independent and state-controlled media have shown an increased willingness to investigate and report on environmental abuses. The rise of social media and microblogging websites, such as WeChat and Weibo, gives ordinary citizens greater access to environmental information and more opportunities to connect with like-minded individuals and organizations than ever before. Thus, with strong evidence of a growing environmental consciousness and assertiveness among the general public, many observers have noted the emergence of a "green civil society" or "green public sphere" in China (Chen 2010; Ho 2001; Yang and Calhoun 2007). They argue that, as the country continues to develop and access to independent media continues to expand, environmental actors will eventually force the government to accept more policy input from below. In essence, they project that China will follow a path similar to that taken by Western democracies in the mid- to late twentieth century.

However, this forecast of bottom-up environmental reform in China is not without its detractors and skeptics. In fact, many scholars, while acknowledging increased environmental concern among the general public, see little evidence that a mass environmental movement could be initiated or sustained in China's current political climate. Protests and mass demonstrations, while dramatic and headline grabbing, are still the exception, not the norm. Most environmental activists still seek compromise and mediation with government authorities, and even those who protest are typically satisfied with financial compensation, rather than whole-scale environmental reform (Steinhardt and Wu 2016). Furthermore, while the country's new environmental laws may look solid on paper, local enforcement tends to be uneven and weak, a problem compounded by China's lack of an independent judiciary (van Rooij 2012). Even today, accurate and unbiased environmental data are difficult to come by, and examples abound of local governments colluding with businesses and media outlets to cover up or suppress the release of potentially harmful information (Pei 2006; Zhou 2010). Microbloggers and investigative reporters who cross the invisible "red line" in environmental reporting are still subject to intimidation, censorship, and arrest; and even the voices of those who escape uncensored are often drowned out by the cacophony of disparate voices and opinions that proliferate on Chinese social media (Jiang 2016; Liebold 2011). Thus, according to the skeptics, while public opinion on environmental issues may be acknowledged and even consulted from time to time, the true ability to shape environmental policy remains firmly in the grip of government officials.

The following section fleshes out this debate over the nature of Chinese environmental state-society relations in further detail. When assessing the arguments of various scholars and researchers, several key questions must be kept in mind. For example, to what degree have individuals and ENGOs been able to carve out an independent role in shaping environmental outcomes? How do various state actors, at both the local and national level, respond to this bottom-up pressure, and to what extent are they likely to make concessions or adopt changes in policy? Is environmental civil society in China a unified force, or is it a disparate and uncoordinated mess of competing actors that can be co-opted and turned against itself? Similarly, in the realm of environmental policy, can the Chinese government be thought of as a single cohesive entity, or is it instead prone to conflicts of interest that can be exploited by crafty environmental actors? These questions are all essential for understanding the nature of environmental state-society relations in China today and their answers will help to determine the outcome of environmental policies in the future.

Measuring Environmental Public Opinion

Before examining the nature of the relationship between the Chinese state and the public, it is first necessary to ascertain the degree to which the Chinese public understands, and is concerned about, environmental issues. Case studies of individual protests and in situ fieldwork offering in-depth analyses of citizen attitudes at the county or township level are certainly valuable. However, because of China's massive size and geographical diversity, as well as its very large population, one must be careful not to generalize local findings to the country as a whole. Thus, large-scale, nationally representative surveys of environmental public opinion in China, though still relatively rare, offer the best chance for understanding how ordinary citizens feel about recent trends in environmental quality and governance. Even here, however, the results are mixed and frequently open to interpretation.

For example, Harris (2006) synthesizes more than a dozen environmental public opinion surveys and concludes that environmental awareness in China is comparatively low and that most ordinary citizens understand little about the scientific principles underlying environmental degradation. Furthermore, he reports that most Chinese express a striking indifference towards environmental protection and are largely content to leave policy decisions to local government officials. This relative lack of concern is echoed by Liu and Mu (2016), who use data from 2008 to conclude that environmental degradation is not generally listed as a top priority for most citizens. Rural residents in western provinces are especially prone to apathy regarding environmental issues, while high levels of environmental concern are mostly limited to wealthier urbanites living in coastal regions (Yu 2014). These findings are challenged by other opinion surveys which show that Chinese citizens of all backgrounds are becoming increasing aware of and interested in environmental issues. Many·scholars also find that Chinese citizens attach a high degree of significance to environmental problems and that public desire to address these problems and take part in policy solutions rivals that in more developed countries (Ding 2016; Liu and Leiserowitz 2009; Saich 2015).

Although this inconsistency in results is puzzling, part of the discrepancy can be explained by differences in the timing and specific wording of survey questionnaires. For the most part, surveys showing very low levels of environmental knowledge and concern were conducted in the late 1990s and early 2000s, but surveys conducted since the mid-2000s show a much higher degree of citizen engagement. Also, the degree to which citizens profess concern about environmental issues varies greatly, depending on *which* type of environmental issue is discussed. In general, the Chinese public expresses greater concern about environmental issues that are easily observed and affect their day-to-day lives (e.g., air and water pollution) and more limited concern about less immediate issues such as global warming and habitat destruction (Horizon Key 2015).

As mentioned earlier, wealthier urbanites and more educated individuals typically exhibit higher levels of environmental concern, a finding mirrored in surveys conducted in the United States and many other countries. However, in addition to individual socioeconomic and demographic characteristics, research shows that macro-level factors also play an important role in shaping environmental attitudes. In their 2016 study, Liu and Mu report that, after county-level fixed effects are accounted for, the significance of individual characteristics declines considerably, suggesting that county-specific economic and political factors cannot be ignored when analyzing environmental attitudes. Local pollution levels may also affect public opinion, as evidenced by studies showing diminished levels of subjective well-being in localities with comparatively high concentrations of air and water pollution (Smyth et al. 2011). This presumed relationship between observed pollution and environmental opinion is understandably complex, with some researchers showing a clear correlation and others producing more ambiguous results. For example,

Dasgupta and Wheeler (1997) find that the link between pollution and perception exists only for easily detected pollutants, such as airborne dust, whereas the link is more tenuous in the case of less-visible SO_2 and COD. Likewise, Silva et al. (2012) report that it is not the absolute concentration of airborne particulates but rather the *change* in these concentrations over time that most affects the attitudes of local residents.

Other surveys focus on the acceptance of risk from specific projects with the potential to cause negative environmental impacts. For example, Huang et al. (2013), in a study of the chemical industry in Jiangsu Province, find that the willingness of citizens to tolerate the risks associated with new chemical plant construction is related to a wide variety of demographic and psychometric factors. The authors report that risk acceptance increases with knowledge about the chemical industry, trust in government, and the perceived economic benefits of the plant's construction. By contrast, risk acceptance decreases with the perceived likelihood and effect of accidents. However, none of these psychometric factors is significant across all demographic groups, and different segments of the population respond differently to potential risks and benefits. For instance, highly educated respondents and men under age fifty are more willing to tolerate risks and focus on the potential economic benefits of new chemical plants, while less-educated respondents and the elderly are more sensitive to the risk of accidents. Notably, risk acceptance is lowest among middle-class residents but increases at both ends of the income spectrum. Together, these results suggest that the nature of the relationship between public opinion and environmental policy outcomes varies widely across China, depending upon which specific demographic and socioeconomic groups are affected.

In addition to measuring environmental awareness, concern, and risk acceptance, many surveys ask respondents to rate their level of satisfaction with government handling of environmental issues. In general, most Chinese citizens are fairly satisfied with the actions of the central government, even if they consider environmental degradation a serious issue (Ding 2016; Saich 2015). Moreover, satisfaction has risen in recent years, and a majority of Chinese citizens express optimism that the central government will be able to improve environmental quality in the future (Liu and Leiserowitz 2009). At the local and provincial level, however, opinions are more mixed. In particular, citizens are skeptical of their local government's ability to monitor and enforce standards for air pollution, water pollution, and food safety (Horizon Key 2015). This stark divide between central and local government satisfaction levels mirrors the results of previous studies on other issue areas (Liu 2015; Saich 2015).

Finally, some surveys ask respondents about their participation in environmental activities (both actual and hypothetical) as well as their willingness to complain or engage in environmental protests. Generally, most Chinese citizens appear willing to sacrifice some economic well-being in terms of lower income or higher taxes in order to obtain a cleaner environment. In fact, this willingness-to-pay (WTP) among Chinese citizens is at least as high as that of citizens in the United States and Europe, adjusted for income levels (Leene et al. 2014; Zhang, Zhang, and Chen 2015). In terms of participating in environmental complaints and lawsuits, a majority of Chinese citizens appear open to the idea, but only a very small percentage has actually engaged in such actions (Chen et al. 2013). Overall, existing opinion surveys depict a Chinese public that is increasingly aware of and concerned about pollution and other environmental issues but that is not yet engaging in acts of mass participation or resistance against the government or polluting industries.

Environmental Complaints, Protests, and Citizen Action

Measuring public opinion is a useful way to "take the temperature" of the general populace, but even widely held environmental beliefs are unlikely to produce any real policy changes unless they lead to some kind of bottom-up action. As outlined above, citizen complaints and environmental protests are no doubt increasing, but the question remains: which factors induce individuals and groups to take action, and, once taken, what determines whether these actions are successful?

Deng and Yang (2013) propose that Chinese citizens distinguish between "insider" and "outsider" pollution, and are more willing to protest when they perceive pollution as originating from outside their own locality. By contrast, when they view pollution as local in origin, they are more likely to tolerate it as a necessary byproduct of economic development. Thus, citizens who are employed by polluting enterprises or have a direct financial stake in their continued profitability are far less likely to challenge their operations or push for stricter environmental regulations (see also Lora-Wainwright 2013; Tilt 2013). Moreover, Lora-Wainwright et al. (2012) argue that even those initially opposed to pollution can become accustomed to it and resigned to its inevitability over time. This phenomenon of "learned helplessness" can arise as citizens experience the futility of trying to change environmental policy at the local level. When complaints are continually ignored by businesses and government officials, and when individual protesters are simply suppressed or "bought off" by local leaders, citizens often adopt an attitude of fatalistic pessimism about continued environmental degradation.

Even when local environmental complaints and protests gain traction, participants are typically satisfied with financial reimbursement for individual damages, an outcome that van Rooij (2012) terms the "compensation trap." For instance, Cross and Snow (2010) report that protesters are more likely to be roused by environmental incidents that have clearly and measurably damaged the health and financial well-being of themselves or their families. Therefore, if their own remonstrances have been properly addressed and compensated, protesters are typically willing to come to terms and withdraw their opposition, even if broader issues of policy have not been dealt with. Similarly, Johnson (2010) finds that most environmental activists exhibit a "not in my backyard" (NIMBY) mentality when it comes to the siting of potentially harmful development projects. Thus, protesters may not have any problems with the proliferation of dirty industries or lax environmental standards per se, so long as those industries are not located in their own city or village. According to Johnson and others, this relatively narrow and self-interested mind-set may produce short-term victories as individual projects are scrapped or relocated in the face of opposition, but it is unlikely to lead to more sweeping, long-term changes in China's environmental policy. Furthermore, local governments may be able to exploit these attitudes by offering generous financial compensation to a small group of protesters in an attempt to divide and conquer (Lora-Wainwright et al. 2012). Alternatively, they may simply decide to relocate environmentally harmful projects to poorer, rural areas, which are less able to effectively resist them (Shapiro 2012).[8]

Although some Chinese environmental activists are no doubt motivated by pure self-interest, some scholars see evidence of a growing popular commitment to the broader ideals of environmental justice and sustainability. In his 2010 study, Johnson finds that, although NIMBYism was rampant, anti-incinerator protesters in Beijing and Guangzhou gradually decided to shift their focus to a more broad-based critique of trash incineration in general. By strategically framing their arguments in terms of more universal environmental ideals, the protesters were able to counteract government claims that they were acting selfishly and not in the broader public interest. Other studies use the New Ecological Paradigm (NEP) scale developed by Dunlap and colleagues (2000) to argue that Chinese citizens increasingly view the environment, not as an external entity to be tamed and exploited but, rather, as a complex web of ecological interdependencies that must be protected from human interference. These claims are bolstered by recent surveys in which a remarkably high proportion of Chinese respondents express a desire for stricter environmental regulations, even at the expense of higher taxes or slower economic growth (Horizon Key 2015; Liu and Leiserowitz 2009). Thus, while apathy and self-interest may still

---

8 For more on how local governments respond to environmental complaints and protests, see Li et al. (2016).

be the norm in China, evidence is increasing that major shifts in environmental public opinion have already begun to take hold.

The propensity to complain or protest may also be driven by environmental and political conditions at the local level. Bretell (2003) and Dasgupta and Wheeler (1997) find that the population-adjusted rate of environmental petitions submitted to government authorities increases proportionately with local concentrations of suspended particulates and SO_2 respectively. By contrast, Zhong and Hwang (2015) report that local air quality is not correlated with people's willingness to engage in environmental protests and that *perceived* pollution levels are only weakly correlated with it. Instead, people's general trust in government and perceptions about corruption are much better predictors of their willingness to protest. Finally, Ding (2017) argues that in China people tend to lend their support to well-publicized, government-initiated environmental actions, regardless of whether those actions actually lead to improved environmental quality. Future research is needed to reconcile this disparity in results, because untangling the relationship among environmental conditions, local politics, and citizen action is of paramount importance in understanding the true relationship between the Chinese state and the general public.

In addition to examining the reasons *why* citizens decide to complain or protest, many studies seek to understand the conditions under which these complaints and protests are most likely to lead to a positive outcome. Mertha (2008) writes that environmental protests tend to be more effective when "policy entrepreneurs" with personal connections to media and government officials take the lead in organizing. Similarly, Deng and Yang (2013) stress the importance of issue framing for environmental activists. For example, when engaging in acts of environmental opposition, protesters may choose to "piggyback" on issues unrelated to pollution, thereby allowing more opportunities to maximize the vulnerability of their opponents and gain public sympathy. Protesters may even play on themes of Chinese nationalism, as the residents of Qidong did in 2012, when they successfully resisted the construction of an industrial wastewater plant by highlighting the company's ties to Japanese business interests and raising the age-old specter of Japanese interference in Chinese affairs (Jian and Chan 2016). According to Jing (2003), local protesters also often draw upon deeply rooted Chinese cultural themes, such as "kinship, popular religion, moral concerns, and ancient tales of justice" (p. 208) to frame their environmental protests. Thus, by consciously expanding the scope and general appeal of their arguments, environmental activists can build a broader and more energized coalition that is more likely to induce government officials to take substantive policy action.

Environmental protesters can also seize upon divisions and conflicts within China's political system itself, playing different agencies and factions off against one another in order to achieve their desired outcomes. In advancing this argument, Mertha (2008, 2009) again points to the fragmented nature of China's bureaucratic system, in which different interest groups and levels of government fight to modify or veto one another's preferred policies. According to Mertha, this pluralistic expression of Chinese political power allowed opponents of hydroelectric dams along the Nu and Min Rivers to marshal support among certain segments of the state-controlled media and environmental bureaucracy, thus providing an effective counterweight to government-backed proponents of the dams. By contrast, protesters along the Dadu River in Hanyuan County failed to mobilize and engage government officials, the media, or the general public, and thus dam construction there proceeded as scheduled.

The Nature of ENGOs and "Green" Civil Society

State-society relations in the environmental sphere cannot be properly understood by simply summing up the various exchanges among individual environmental actors. Rather, it is necessary to take a more holistic view of China's complex web of interpersonal associations, both inside and outside the government. Therefore, this section offers a fuller analysis of China's ENGOs and the "green" civil society in which they operate.

Of course, some observers question whether it is accurate to presume that a green civil society exists in China at all. White, Howell, and Shang (1996) define civil society as an autonomous sphere of voluntary associations with the potential to organize the interests of socioeconomic groups that are developing as a consequence of reform and might be able to counterbalance the unchallenged dominance of the state. However, the authors find little evidence of a clear separation between state and society in China, on environmental issues or otherwise. Instead, they report a more complex intermingling between the Chinese government and various societal actors, with relationships most often taking the form of mutual dependence and jurisdictional overlap. With this in mind, Frolic (1997) and others have coined the term "state-led civil society," cautioning that researchers should not equate China's unique situation with the forms of civil society more traditionally found in the West. Focusing on the environment, Zhang and Barr (2013) reach a similar conclusion, citing the lack of organizational capacity among ENGOs, the continuation of government restrictions on ENGO formation and fundraising, and the tendency of citizens to rely on the government to solve their environmental problems.

Other researchers, in contrast, find evidence of a rapidly burgeoning green civil society in China. For example, Yang and Calhoun (2007) argue that changing mind-sets and modes of communication among environmental activists have allowed the rise of a Chinese "green public sphere." Xie (2015) also reports that ideas of "thick citizenship" and "greenspeak" have spread among the general populace, creating a network of environmental organizations that are increasingly willing to operate independently of and—if need be—in opposition to the government. Enlightened citizens and ENGOs can even use the language of China's own environmental laws to assert their rights, a phenomenon noted by O'Brien and Li (2006), Wang (2014), and Yang (2005).

In some sense, of course, the debate over the existence of a Chinese green civil society is irrelevant. The fact remains that various government, quasi-government, and independent groups do exist in China, and they must interact with one another when addressing environmental issues. Thus the essential question is: what is the exact nature of these interactions and to what extent are they characterized by expressions of dominance, submission, cooperation, opposition, and everything in between?

ENGOs in China have existed since the early 1990s, but recently they have greatly increased their clout within China's political sphere. Between 2005 and 2008 alone, the number of ENGOs in China doubled (Zhan and Tang 2013), and since then they have continued to increase in importance and capacity (Hildebrandt 2013; Kostka and Mol 2013; Zhan and Tang 2013). As Chinese politics grows more pluralistic, the government's treatment of ENGOs has shifted from a policy of exclusion to one of acknowledgment and incorporation. According to Steinhardt and Wu (2016), ENGOs have likewise begun to alter their narrow and "cellular" approach to politics and embrace a more universal form of "popular contention."

However, the power of Chinese ENGOs is still tightly circumscribed. Officially, all NGOs must be approved by, and registered with, the Ministry of Civil Affairs (Zhan and Tang 2013), and, as of 2016, foreign organizations must also register with the police (Wong 2016). Chinese NGOs are prohibited from engaging in public fundraising or setting up branch organizations in other countries or provinces (Shieh and Deng 2011). Moreover, a high proportion of NGOs are separate from government in name only. In reality, these government-organized NGOs (GONGOS) (Schwartz 2004), are staffed by state appointees and may even receive state salaries, thus making it nearly impossible for them to challenge official policies. Even when they are independent, most Chinese ENGOs find it difficult to operate and are typically faced with a shortage of funds, information, and qualified staff (van Rooij and Lo 2010). Therefore, many observers caution that, although they may be growing in number and importance, the vast majority of ENGOs in China are still characterized by fragmentation, disorganization, and reliance upon the state (Ho and Edmonds 2007; Zhan and Tang 2013).

The enhanced prominence of Chinese ENGOs can be attributed in part to a decline in the effectiveness of formal environmental institutions. As the state retreats from its role as the sole provider of services and gatherer of information, civil society can step in and fill the gap. Moreover, because local EPBs often face shortages in funding and staff, ENGOs can help to lighten the load of government burdens (Holdaway 2013; Zhan and Tang 2013). Thus, in some ways, the state is just as reliant on ENGOs as ENGOs are on it.

After ENGOs are established, they must decide how to best negotiate their delicate relationship with the various levels of the Chinese government. According to Ho (2001, 2007) and others, most choose to adopt a nonconfrontational stance vis-à-vis the state, often engaging in acts of "self-censorship" to avoid antagonizing government officials. Some groups go even further and pursue a strategy of "voluntary co-optation," embedding themselves within the state and working as closely with government officials as possible. Although some regard this as a sign of weakness, Baum and Shevchenko (1999) note that it may also be considered a pragmatic and strategic act, designed to enhance their standing in the eyes of government officials and eventually gain political influence. In the same vein, Xie (2015) explains how the leaders of many Chinese ENGOs rely heavily on acts of "networking" and "bridge building," cultivating personal connections and embedding themselves within the Party-state and the media. These connections can pay off down the road, causing government officials to be more sympathetic to environmental demands and grievances from the public. Xie (2011) also notes the contrast with Russian ENGOs, which have high levels of sophistication and development but lack personal connections with state and media leaders, thus limiting their effectiveness.

Although environmental state-society relations in China have been marked by increasing openness, the government remains the dominant actor. Different observers, however, disagree about the extent to which ENGOs are independent of, or reliant upon, the Chinese state. They present arguments that run the gamut, from those that see complete state control to those that find emerging ENGO independence, but most settle on a middle ground. The following offers a brief summary of some of the most prominent models of state-ENGO relations in China, ranging from the most to the least state centric.

In his article on "authoritarian environmentalism," Gilley (2012) writes that, in China, environmental decision-making is "limited to a narrow cadre of scientific and technocratic elites while others are expected to participate only in state-led mobilization for the purposes of implementation" (p. 288). In other words, ENGOs and other non-state actors only to assist and obey the central government and can continue to operate only as long as they remain in its good graces. A somewhat less stark version of state dominance is advanced by He and Warren (2011) and Weller (2008, 2012).[9] Essentially, the government tolerates and even encourages participation and input from non-state environmental actors but, at the same time, maintains a firm and unyielding grip on the power to make all final environmental policy decisions. This view is similar to Frolic's (1997) previously mentioned concept of "state-led civil society," in which NGOs operate in cooperation with, rather than in opposition to, the government.

Taking a slightly different view, some scholars argue that the relationship between Chinese society and the state can best be understood as a form of "corporatism" (Hsu and Hasmath 2014; Unger and Chan 1995). In the words of Zhang and Barr (2013):

> The state establishes itself as the arbitrator of legitimacy by limiting and licensing the number of players with which it must negotiate its policies, essentially co-opting their leadership into policing their own members. (p. 11)

By granting legality and a degree of independence to a carefully-selected group of ENGOs, the government can withdraw its control over certain environmental affairs without having to worry about undue social unrest or disorder. In a sense, this theory lines up with the idea of "mutual empowerment" put forth by Gu (2011), in which society and the state rely on each other. Society depends on the state for legitimacy and resources, while the state depends on society for input and information. Finally, other researchers find that some Chinese localities have taken tentative steps toward "consultative authoritarianism" (Teets 2014) or even "deliberative democracy" (Fishkin et al. 2010), in which government officials seek input (either formal or informal) from below and shape their policies to better reflect the will of the people.

Of course, the possibility exists that more than one of these conceptual relationships exists simultaneously in different contexts. For example, Spires (2011) argues that state-society relations in China are marked by a "contingent symbiosis." Grassroots NGOs can flourish and enjoy relative autonomy as long as they render valuable services to the government and refrain from "democratic claims making" that might fuel grievances toward the state and lead to

---

9   He and Warren (2011) use the term "authoritarian deliberation," while Weller (2012) uses the term "responsive authoritarianism."

social disorder. Thus, China's level of grassroots autonomy is entirely dependent upon how closely NGOs remain within the prescribed political boundaries delineated by the state. Making a similar assertion, Ho and Edmonds (2007) put forth the concept of "embedded activism," arguing that civil society can be quite vocal and active in China, as long as it avoids directly challenging the legitimacy of the central government.

In addition to exploring the ways in which Chinese ENGOs operate and interact with the state, it is also necessary to examine how government actors respond to entreaties and challenges from the public. Martens (2006) and Xie (2011) write that ENGOs can actually be of great help to government agencies by helping to collect data and monitor potential lawbreakers. Local governments in particular, which typically suffer from a lack of administrative resources, often find that they can take credit for ENGO successes and simply shift the blame to them if things go wrong (Spires 2011). Thus, as the problems of environmental governance become more complex and the financial resources for dealing with them grow more limited, government officials have consciously reduced their restrictions on ENGO operations, allowing them greater input in the governing process.

However, greater input does not necessarily mean enhanced decision-making power. Stern (2013) argues that, although most government officials want to protect the environment and avoid the social instability that widespread pollution causes, they are reluctant to sacrifice their primary mission of promoting economic growth and maintaining political control. By trying to have it both ways, their attitude toward Chinese ENGOs is often contradictory and marked by mixed signals, a phenomenon she calls "political ambivalence." Presenting a similar viewpoint, Carter and Mol (2006, p. 336) write that the Party-state welcomes support and assistance from ENGOs but not their efforts to move toward a more "active, democratic society." Therefore, increased ENGO participation and autonomy is a double-edged sword for government officials, tolerable in small doses but potentially dangerous in its longer-term implications. Some government agencies attempt to walk this fine line by varying their level of control according to context-driven factors, such as geographic location, the types of issues that ENGOs focus on, and the specific local officials and agencies involved. By participating in this regime of "tacit sanctioning," ENGOs can develop long-term relationships with their government counterparts and gradually gain an understanding of government needs and the political boundaries implicit in their operation (Hsu and Hasmath 2014; Spires 2011).

Other authors are careful to note that the Chinese state cannot always be perceived as a single, unified entity. Indeed, as mentioned earlier, fragmentation and conflicts of interest within the government can serve as strategic avenues for increased ENGO influence. Zhang and Barr (2013) note that, within local government, EPB officials tend to be more progressive and tolerant of ENGO operation, while local Party secretaries with connections to local businesses tend to be more investment driven and skeptical of environmental groups. Official sympathy for environmental complaints and protests also tends to be concentrated in certain sectors of the government more than others, as evidenced by the uneven official response to Chai Jing's inflammatory *Under the Dome* documentary in 2015.[10] Initially, many government officials declared their support for the film, only to be overruled by more conservative elements within the Party-state a week later (Larson 2015). Thus, because of divisions within the government, the official response to environmental challenges tends to be ad hoc and inconsistent, and there is no institutionalized or commonly agreed upon blueprint for engaging with the public on environmental issues (Li, Liu, and Li 2012).

In some instances, individual state entities reject this passive approach to bottom-up pressure and take the lead in creating new channels for grassroots participation. Zhang (2010) writes about Fuyang's Green Bounty Hunter Campaign, in which local EPB officials offered citizens financial rewards for monitoring polluting firms and reporting environmental violations to the government. Government agencies have also responded to pressure from below by creating their own ENGOs and sponsoring their own surveys to gather environmental opinions from the public (He and Thøgersen 2010; Schwartz 2004; Shapiro 2012). However, Grano and Zhang (2016) caution that some of these actions may simply create "decoy" participatory channels, designed to placate citizens without allowing them the chance to offer any meaningful input. Similarly, Wang (2013, p. 434) argues that, for the most part, "public supervision channels have been more about government information gathering than true problem solving on behalf of the public."

In the end, government officials tend to enlist citizens' help in addressing environmental challenges while also directing their efforts through "institutional channels" in a way that does not jeopardize state legitimacy (Grano and Zhang 2016). However, some researchers acknowledge that this is a dangerous game. Increased participation and dissemination of information, no matter how closely controlled, can gradually foster a sense of political efficacy and shared identity, sowing the seeds of a civil society that will eventually seek empowerment (Xie 2016). Thus, in the long run, the Chinese government may

---

10   For more on "Under the Dome" and the government's reaction to it, see Wong (2015) and
       Wildau (2015).

end up ceding ever-greater amounts of environmental decision-making power to the public—not because it wishes to but because it must.

The Role of the Media in Environmental State-Society Relations

Any discussion of Chinese state-society relations must also acknowledge the mediating and moderating role played by the country's fourth estate. In particular, scholars have paid close attention to the rapid rise of environmental activism on the internet, to which more than 700 million Chinese now have access (CNNIC 2016). With its advantages in speed, interactivity, and breadth of dissemination, Yang (2009) and Zhang (2016) argue that the internet has facilitated the rise of a green public sphere in China. Indeed, survey work conducted by Wang and Cheng (2015) shows that increased internet use among Chinese citizens is associated with higher rates of environmental activism, while increased television use produces the opposite results. In assessing these trends, some observers predict that the internet will eventually lead to the emergence of a more democratic civil society in China as new classes of netizens cleverly avoid government censorship and rally the Chinese public (Xiao 2011).

However, others are less sanguine about the prospects of an internet-led democratic revolution in China. Jiang (2016) argues that the internet, far from uniting public opinion, can actually serve to polarize and segregate users into their own individual echo chambers. People may also become sidetracked by the triviality and banality of popular online news, preferring escapism over action and turning the internet into a form of "digital opium" (Liebold 2011). Kay, Zhao, and Sui (2015) report that access to the internet in China is still highly uneven, with older, poorer, and more rural populations largely excluded. This poses significant problems for online green activists, because cooperation and support from these marginalized populations are essential if bottom-up pressures are to succeed.

Of course, the Chinese media landscape is not limited to the internet, and some evidence indicates that other, more traditional forms of news media are also playing a role in China's environmental movement. Yang and Calhoun (2007) write that environmental reporting in national and regional newspapers is rising and that mass media outlets are becoming more commercialized and "de-ideologized," giving them greater latitude in covering acts of environmental unrest. Other scholars point to the rise of environmental investigative journalism as evidence that the Chinese media are becoming more critical of the government and less willing to respect traditional political boundaries (Tong 2015). In a similar vein, Tilt's (2007) case study on industrial pollution in rural Sichuan Province demonstrates how the media can be a powerful agent in exposing pollution and improving policy enforcement.

However, even optimists concede that true freedom of the press is not a reality in China. For the most part, Chinese print and television media take pains not to challenge the government's legitimacy, and although they may criticize government handling of specific environmental incidents, they typically refrain from condemning the system as a whole (Fedorenko and Sun 2016; Tang and Tang 2013). Moreover, van Rooij (2010) writes that, although Chinese news media are generally given free rein in criticizing local leaders, disparagement of national-level officials is still strictly verboten. Nevertheless, the situation remains more relaxed than in the past. For example, Tilt and Xiao (2010) report that even state-controlled media outlets have sometimes shown a willingness to push boundaries, as CCTV (China Central Television) did when it exposed government cover-ups in 2005 after a benzene spill in the Songhua River.

Although environmental reporting in China is still in its infancy, green activists have already used their media savvy to induce real changes in official policy. Among these grassroots success stories, perhaps the best known is the campaign to encourage government monitoring and publication of data on real-time air quality. In 2011, after the US government installed an air-quality monitor on the roof of its embassy in Beijing and began publishing the results on Twitter, many Chinese citizens were shocked and outraged to learn that urban air quality was far worse than government officials had led them to believe (Zhang 2016). In response, Chinese activists took to Sina Weibo, a microblogging platform that reaches an audience of 400 million, demanding that central leaders adopt mandatory standards for monitoring particulate matter with a diameter of less than 2.5 micrometers ($PM_{2.5}$) and other pollutants. Just a few weeks later, the MEP announced that it was officially revising China's National Ambient Air Quality standards and would make hourly air pollution data available to the public, a victory largely made possible by the remarkable scale and coordination of the online protests (Fedorenko and Sun 2016; Huang 2015). In other instances, government officials have even encouraged these types of media campaigns, and Tang and Tang (2013) write that grassroots reporting has often prompted local governments to crack down on pollution infractions that they otherwise would have missed. Indeed, Zhang et al. (2010) argue that, in general, Chinese firms tend to be more nervous about hostile media reports than EPB fines or sanctions. Thus, transparency-based ENGOs have been able to achieve real policy influence through their ability to "name and shame" local polluters, an ability enhanced by their often close personal ties with independent and state-owned media outlets (Xie 2009, 2011).

Despite these close ties, Party officials and government-controlled media are not always so amenable to challenges from below. Van Rooij (2010) writes that state-owned media often characterize environmental protesters and ENGO leaders as unruly or mentally ill, while other researchers note that the Chinese government has *increased* its control over the internet in recent years (Zhan and Tang 2013; Zhang 2016). Government agencies do not generally censor individual complaints or isolated criticism of state policies, but any complaints that gather momentum and threaten to undermine Party legitimacy are quickly targeted and erased (King, Pan, and Roberts 2013). Government officials may even use social media for their own purposes, by intercepting text messages or monitoring Weibo to catch wind of planned protests and dispatching police forces accordingly (Kay, Zhao, and Sui 2015). Finally, state authorities and corporations may "co-opt" media platforms in order to advance their own agendas, engaging in public relations campaigns and pushing back against the efforts of environmental groups. In acknowledging this continued government control over new media, MacKinnon (2012) suggests the emergence of a "networked authoritarianism," warning that, by itself, the internet is not a driver of political change. Rather, it is simply a tool—like any other form of media—and it can be exploited by activists and government alike.

Formal Complaint Procedures, Legal Challenges, and the Judiciary

In addition to engaging in protests and spreading information through the media, environmental activists may also attempt to induce policy changes through formal institutional channels. For individuals, the most common means of formal action is to lodge an environmental complaint or petition. Complaints can be issued online, by phone, in person, or in writing; and, according to van Rooij et al. (2016), total annual environmental complaints in China rose from 268,592 in 1999 to 735,756 in 2010.[11] The vast majority of these complaints receive at least some sort of formal government response (Dong et al. 2011), and some evidence shows that the population-adjusted rate of environmental complaints in a region is positively correlated with EPB fines and sanctions against illegal polluters (van Rooij and Lo 2010). However, other scholars note that environmental complaints and petitions originate disproportionately from wealthier, more educated regions, suggesting that marginalized populations may not be receiving the same level of positive benefits (Dong et al. 2011). Citizens who lodge formal complaints tend to focus on "nuisance problems" that are personally bothersome, such as noise pollution, while ignoring other environmental problems that are less easily observed (van Rooij 2010; Warwick and Ortolano, 2007; Yang and Zhang 2012). Finally, local officials have a strong incentive to repress environmental complaints and petitions

---

11   After 2010, the government stopped reporting data on the annual number of environmental complaints.

that reflect badly on their governing abilities, thus increasing the rate of conflict between local governments and their citizens (Wong and Peng 2015).

In addition to lodging formal complaints and petitions, individuals can enlist the help of ENGOs to push back against corporations and government officials by taking them to court. Environmental courts are a relatively new phenomenon in China, and their numbers have grown rapidly, with 130 opening between 2007 and 2013 alone (Stern 2014). In fact, some Chinese ENGOs are now devoted solely to providing legal aid and assistance for pollution victims (Schwartz 2004).[12] In the eyes of some observers, the rise of China's environmental courts has already played a major role in altering state-society relations. According to Zhang (2008), environmental courts have enhanced the power of local and provincial EPBs, allowing them to enforce environmental penalties and transparency laws that were previously not taken seriously. In addition, Yong (2009) argues that courts can help to dampen protests and reduce social conflict by arbitrating environmental disputes to the satisfaction of both parties. Also, while she is careful to note their current lack of enforcement power, Stern (2014) writes that environmental courts can help to build state capacity and strengthen civil society by offering a degree of legal institutionalization and serving as "consciousness-raising institutions" for the general public. Moreover, evidence indicates that, unlike China's complaints and petitions system, access to environmental courts is fairly equitable across class and geography, with the majority of cases initiated by rural farmers in poorer regions (van Rooij 2010). Thus, while China's environmental courts are undeniably flawed and not yet fully developed, their growth has increasingly inspired confidence among activists hoping to alter the balance of power between ordinary citizens and the Party-state.

Other researchers, however, are far more skeptical about the effectiveness of China's legal system in resolving environmental issues. For example, Kennedy (2012) reports that less than 1 percent of all environmental disputes are resolved in court, with criminal lawsuits being especially rare (Economy 2014). Even when the courts intervene, punishments against violators tend to be weak, and fines are often reduced after the fact by sympathetic state officials (Liebman 2007; Zhang 2014). Well-connected firms are particularly immune to legal action (Lorentzen, Landry, and Yasuda 2014), and ENGO challenges against government agencies are even less likely to succeed (Jin 2015). In fact, according to

---

12   The best-known is the Center for Legal Assistance to Pollution Victims (CLPV), which was founded in 1999 and is based at the Beijing University of Law and Politics. The CLPV seeks to train environmental law enforcement and legal officials, improve public awareness of environmental laws, and provide legal services to plaintiffs in environmental court cases.

China's new environmental laws, individual citizens are still legally prohibited from initiating lawsuits on their own, and among the country's thousands of ENGOs, only the 300 largest are permitted to bring cases to trial (Hsu 2014).

Unsurprisingly, these facts have exposed China's environmental courts to accusations of bias. For example, Su and He (2010, p. 181) argue that China's lack of judicial independence means that judges must be thought of as agents of the state, writing that "Chinese courts are a branch of local government ... tightly meshed with other departments and subject to Party leadership." Furthermore, because courts are often punished for high rates of retrials, appeals, or petitions, they typically shy away from environmental disputes that are not easily mediated (Liebman 2011; Su and He 2010; van Rooij, Stern, and Fürst 2014). They also tend to avoid hearing cases that involve powerful business leaders with a high degree of political clout (van Rooij, Stern, and Fürst 2014). Finally, administrative hurdles within the legal system may erode the will and solidarity of participating citizens by subjecting them to interminable delays and red tape, inducing feelings of "informed disenchantment" and "learned helplessness" (Gallagher 2007; Lora-Wainwright et al. 2012).

With all this in mind, skeptics argue that China's environmental courts may offer occasional legal victories and policy successes but that true environmental reform will be impossible until the country's legal system achieves full independence from the CCP. Indeed, the same can be said for all the forms of environmental state-society relations detailed above. The Communist Party holds the keys to all final decision-making power in China, whether with respect to NGO activism, social protests, media coverage, or legal intervention. Although the government can no longer control all aspects of everyday life, and although the state has ceded a significant degree of influence to societal groups, the CCP retains the ability to reassert control over issues it considers to be of top priority (Edin 2003; Kennedy 2007). Thus while state-society relations are certainly more complex and pluralistic than in the past, the state unquestionably retains immense power to shape China's environmental policies.

4   Market Forces and the Environment

Underlying this relationship between government actors and society at large is the more subtle and difficult-to-define power of the market. Since the beginning of the reform era, scholars have debated the extent to which classical ideas about the free market can be applied to China's nominally communist state. However, despite the continued presence of strict government oversight, the role of market mechanisms in Chinese policy-development has undoubtedly expanded significantly. This section focuses on examining how market forces—broadly defined—shape China's environmental conditions. More specifically, it provides a systematic literature review of market-based environmental governance measures (e.g., cap-and-trade markets and payment for ecosystem services), environmental information disclosure practices (due to their influence on firm and consumer behavior), and the rise of green finance.

Market-Based Environmental Governance

Since the late 1980s, a gradual shift has occurred from a command-and-control style of environmental regulation—the type characterized by technological mandates—to an adoption of more market-based policy tools. When the environmental sector is governed through market mechanisms, the basic tenet is the creation of economic incentives for specific policy actions. Examples includes putting a price on pollution through Pigouvian taxes or cap-and-trade programs, as well as providing compensation for maintaining ecosystem services. This section briefly reviews three examples of market-based environmental governance schemes that have been implemented in China—the SO_2 trading program, carbon markets, and the Grain for Green program.

Emissions Trading Programs

Over the past three decades, China has initiated several waves of policy experimentation as part of an effort to curb SO_2 emissions using pollution trading systems. The first phase of emissions trading development occurred between 1990 and 1994, during which pilot trading programs were established in six cities. Later, during the tenth FYP period (2001–2005), SEPA inaugurated the 4 + 3 + 1 program by selecting four provinces, three cities, and one company to take part in a pilot program for SO_2 emissions trading, while also imposing a hard cap on emissions, known as total emission control (TEC). In the subsequent eleventh FYP period (2006–2010), the Chinese government announced a 10 percent reduction target for two key pollutant criteria, SO_2 and COD, and established additional pilot programs for emissions trading in several provinces (Zhang et al. 2016).

Despite these innovations, however, SO_2 trading in China remains at the pilot stage and has never been implemented on a full, nationwide scale. Even the operational effectiveness of existing programs has been questionable. For instance, Zhang et al. (2013) find that, although China was able to achieve its SO_2 reduction goals during the eleventh FYP period, this success was mainly due to government subsidies for de-sulfurizing equipment in power plants rather than pilot SO_2 trading programs. Zhang and his colleagues also found that the majority of the programs' cost-saving potential was offset by the existence of other overlapping policies.

Tao and Mah (2009) have likewise called into question the market effectiveness of China's SO_2 trading programs. Based on their interviews and site visits to a few selected programs during 2004 and 2005, they found that trading markets were very thin—with only small number of participants and transactions. In what was ostensibly a market-based policy, the government ending up playing a significant role as matchmaker, and many transactions were brokered by local EPBs, with prices negotiated by government officials. In an ideal market, buyers and sellers of emissions allowances engage with each other continuously in conditions of high liquidity and low transaction costs. However, according to Tao and Mah, these early Chinese SO_2 trading programs were anything but ideal. In essence, the gap between theory and practice resulted from the conflicting visions of state control and market competition expressed throughout China's period of economic liberalization—in other words, the Chinese state was still learning to govern through market mechanisms.

Shin (2013) characterizes existing China's SO_2 trading programs as failed policy innovations that were "constrained by domestic institutional factors and the lack of domestic preconditions for effective diffusion and innovation, [which] makes policy adoption costly" (p.920). Shin also points out that local environmental authorities frequently lack the "infrastructure, resources, transparency, and sometimes even willingness" to enforce official policies (928). A more fundamental challenge, highlighted by both Shin (2013) and Tao and Mah (2009) is the nature of central-local relations in China's environmental governance system. As mentioned earlier, local EPBs exercise "dual allegiance" and must report to both EPBs one level above them as well as local governments that provide them with funding, personnel, and other resources. Because local governments control the purse strings of EPBs, and because they tend to focus more on economic development and GDP growth, SO_2 trading programs have generally not received strong institutional support, despite clear enthusiasm from central environmental officials.

Finally, in the most up-to-date evaluation of China's SO_2 trading programs, Zhang et al. (2016) reach a similarly pessimistic conclusion. During 2012 and 2013, they conducted interviews in five pilot SO_2 trading programs and found that the markets were generally thin, congested, and unsafe. The local governments involved tended to be overly active, unnecessarily overcrowding the markets. Participating firms were also troubled by regulatory uncertainty and the tendency of government actors to abruptly alter market rules. Thus, the firms tried to meet several different overlapping environmental regulations at once, creating confusing signals for the market. Together, these challenges constrained the operation of China's SO_2 trading programs and prevented them from being as effective as initially conceived.

Carbon Markets

The problem of climate change has induced the Chinese government to develop trading systems to decrease greenhouse gas emissions. These new carbon trading schemes were drawn in part from China's experiences with the United Nations' Clean Development Mechanism (CDM), in which rich countries invest in low-carbon projects in developing countries to claim carbon credits. Under the direction of the NDRC, China participated in more CDM projects than all other developing countries combined (CDM Pipeline 2016). Thus the country's involvement in international carbon finance markets paved the way for China to adopt its own markets to control domestic emissions.

In the case of Chinese carbon markets, policy development was extremely rapid. Carbon emissions trading was featured in China's FYP for the first time in 2010. The following year, the NDRC approved seven pilot programs in five cities and two provinces, and, by 2014, all the programs had become operational. In designing the pilot programs, the NDRC intentionally authorized each program to operate under its own distinct rules, thus allowing the government to compare the results of different pilot regions and learn from their experiences (Yu and Lo 2015). Finally, after less than two years of pilot program experimentation, the central government announced its intention to implement a nationwide carbon trading system in 2017.

Carbon markets differ from SO_2 trading programs in that they attempt to address a global environmental challenge—climate change—rather than a more local environmental problem. Consequently, the political considerations involved are international in scope. In this sense, Lo (2015) argues that China's carbon market development should be understood within the larger geopolitical context of government efforts to seek power in the global climate governance regime. Lo also notes that the construction of carbon markets has been led almost exclusively by the Chinese state, while private finance has not yet been effectively mobilized. This is a departure from similar experiences in the West, where the majority of emissions trading programs are driven by business-led coalitions. Therefore, China's case is, to an extent, at odds with the argument that market-based environmental programs necessarily exist to serve corporate interests (Lo and Howes 2013).

Although China's carbon markets are still in the pilot phase, many researchers have already launched detailed policy analyses looking at the promises and challenges of the country's existing trading schemes (Jiang et al. 2016; Jotzo and Löschel 2014; Lo 2016; Munnings et al. 2016; Teng et al.

2014; Zhang 2015; Zhang et al. 2014). Several scholars identify the poor quality of emissions data in China as a crucial problem that risks compromising the effectiveness of carbon trading policies. For example, Guan et al. (2012) discovered a 1.4-gigaton gap between China's national and provincial energy statistics. The same research team also found that emissions from domestic cement and fossil fuel production were overreported by 14 percent (Liu et al. 2015). Because of rampant data manipulation and falsification (Wallace 2016), China faces an enormous challenge in establishing a robust monitoring, reporting, and verification (MRV) system to support existing and future carbon markets.

Many scholars have also noted emerging conflicts between market mechanisms and heavy state control in China's power generation and distribution sector. In China, the price of electricity dispatched to firms and households is heavily regulated by the government, and power companies are prevented from passing on any increased generation costs to end users. As a result, carbon prices in China tend to reflect political judgments rather than the true marginal cost of production—a form of price control that undermines the economic efficiency of the market approach (Lo 2016). More generally, Teng and his colleagues (2014) warn that future carbon markets will likely be constrained by China's current overregulation of the power sector and that, in order to address this issue, liberalization of the power sector must be part of the overall policy package guiding China's low carbon transition.

As in China's experience with SO_2 trading programs, scholars have found that domestic pilot carbon markets also typically suffer from low liquidity. Munnings and his colleagues (2016) note that regulated firms face a steep learning curve as they acquire the expertise necessary to participate in these new markets, which limits the amount of initial transactions. Furthermore, the trading volume in pilot programs typically spikes just before the compliance deadline, implying that many firms see the market merely as compliance obligation, rather than as an integral part of their business strategy. The lack of involvement from private finance also contributes to low liquidity (Lo 2016). Finally, as in SO_2 markets, overlapping policies send inconsistent signals to firms that participate in carbon trading schemes, further hindering market participation (Zhang et al. 2014).

Clearly, the Chinese government still needs to address many issues before it can establish its proposed national carbon market. State leaders must determine how to set the overall cap, how to allocate emissions allowances, how to balance regional and sectoral interests, and how to build a robust market infrastructure. The solutions to these problems remain to be seen, but the ultimate effectiveness of the system will hinge upon the degree to which the Chinese government is able to resolve the tensions between the ideology of state control and the preconditions of market competition. Future scholarship would benefit from a deeper analysis into how these markets came into being in the first place, because the existing literature—which is currently dominated by economics and management perspectives—falls short of identifying the political processes that lead to these policies.

Payment for Ecosystem Services: The Grain for Green Program

In addition to adopting emissions trading schemes to deal with pollution, China has established large-scale payment for ecosystem services (PES) plans. Of these efforts, the Grain for Green program (*Tui geng huan lin* 退耕还林), also known as Farmland to Forest and Sloping Land Conversion, is the most notable example (Bennett 2008; Liu et al. 2008). The Grain for Grain initiative sets aside sloped farmland—with payments deliverable in cash or grain—in order to protect forests that surround the headwaters of the Yangtze and Yellow Rivers. Beginning with a pilot study in 1999, the program was expanded to include twenty-five provinces in 2002, and, by the end of 2005, more than RMB 90 billion (roughly USD 11 billion) had been invested in the project. Although the Grain to Green initiative was partially halted in 2007 because of concerns over food security, subsidies for already-converted croplands continued into the 2010s. According to official statistics, by the end of 2013, more than 100 million farmers had participated in the program, with roughly 25.8 million hectares of croplands successfully converted into forest (Feng and Xu 2015). As of early 2017, the Chinese government continued to administer an updated version of the Grain for Green program, with a stated goal of achieving both ecological restoration and poverty alleviation.

Although no national-level assessments of the ecological impacts of China's Grain for Green program have been done to date, researchers generally agree that the impacts on soil and water have mostly been positive (Delang and Yuan 2015). For example, Lu et al. (2012) conducted a case study in the Loess Plateau, finding evidence of significant conversion from farmland to woodland, which in turn led to positive changes in soil conservation, carbon emission reduction, and grain production. Similarly, Zhou et al. (2012) documented a 21.4 percent increase in forest coverage in the same region. However, although several studies argue that China's Grain for Green program has resulted in large-scale carbon sequestration benefits (Deng et al. 2014; Song et al. 2014), others have critiqued the program for being too focused on trees and yielding negative outcomes in grassland ecosystems in arid and semiarid areas (Li 2016). Recent research results also illustrate the tremendous variation in program outcomes as well as the tenuous connection between program implementation and changes in actual vegetation cover, as other local land use practices also play an important role in shaping landscapes (Zhang et al. 2017; Zinda et al. 2017).

Compared to these mostly positive ecological outcomes, the socioeconomic impacts of the China's Grain for Grain program are more mixed. Many studies focus on the issue of mistargeting—a problem that occurs when the program is carried out in locations where implementation is most convenient, rather than in locations with the most acute needs. Thus target areas are often chosen with little regard for land productivity or environmental heterogeneity (Uchida et al. 2005; Wang et al. 2007; Xu et al. 2010) The goal of poverty alleviation has also suffered from systemic implementation problems. Liang et al. (2012) found that the impact of the program varied across sociodemographic groups and that the program did not target poor households, nor did it shift on-farm labor to other types of off-farm jobs. Furthermore, in a case study conducted in Northeast China, close to 60 percent of surveyed households considered themselves worse off after participating in the program, with the majority viewing the program as a form of government coercion (Wang and Maclaren 2012).

Because the Grain for Green program's direct effect on income is often insignificant (Lin and Yao 2014), another thread of the literature addresses whether the program is able to help farmers obtain more non-agricultural work and increase their annual income. The results of these studies are mixed, ranging from positive (Liu et al. 2010; Uchida et al. 2007, 2009; Yao et al. 2010), to neutral or even negative effects (Wang et al. 2007). Finally, with respect to the debate over whether these conservation efforts threaten food security, Xu et al. (2006) find that the Grain for Green program has "only a small effect on China's grain production and almost no effect on prices or food imports." (p.130)

In a 2013 review, Yeh points out that most existing studies on China's Grain for Green program approach the issue from a heavily managerial and technocratic perspective. To a certain degree, the same observation can be applied to research on China's SO_2 trading programs and carbon markets. Such perspectives often assume that a transition to more market-based environmental governance is a desired and inevitable outcome. In fact, all these programs are inherently political, and decisions about the precise control, allocation, and use of resources remain unsettled. Thus, in Yeh's opinion, more careful research into the political dimensions of these programs is needed.

Information Disclosure and Environmental Transparency

In recent years, transparency and accountability initiatives have emerged as an important policy tool for addressing developmental failures and improving democratic governance (Gaventa and McGee 2013). Many scholars have studied the effects of information disclosure as a form of environmental governance

(Gupta 2010; Haufer 2010; Mason 2008), with much attention being focused on China's burgeoning system of regulations on environmental transparency. This section takes a broad view of information disclosure, including policies adopted by the government as well as initiatives originating in the corporate sector. Corporate initiatives include the voluntary disclosure of environmental performance data as well as product labeling to help consumers engage in green consumption.

China's 1979 Environmental Protection Law was the first piece of legislation to provide a legal basis for environmental information disclosure. Over the following decades, state officials gradually expanded and institutionalized environmental transparency policies, culminating in the 2007 Measures on Open Environmental Information (OEI [Huanjing xinxi gongkai banfa 环境信息公开办法]). These regulations mandate disclosure reports on six main areas of information: (1) environmental laws, regulations, and standards; (2) allocation of emissions quotas and permits; (3) pollution fees and penalties collected; (4) exemptions, reductions, or postponements granted; (5) outcomes of investigations into public complaints; and (6) violators of existing environmental regulations.

By ensuring the availability of environmental information, the MEP hopes to strengthen its regulatory power and facilitate public participation (Tan 2014). So far, however, the implementation of China's OEI regulations has not been satisfactory. According to researchers, disclosure tends to be slow and incomplete, and crucial information is often blocked in the name of state security and social stability. In some cases, local officials simply ignore disclosure requirements (Mol et al. 2011; Zhang et al. 2010, 2016). Thus, nearly a decade after the enactment of OEI, local environmental data are still inconsistent and hard to come by (Hsu 2012).

Although implementation remains spotty, environmental transparency measures have had some concrete benefits. For example, Haddad (2015) writes that the Institute of Public and Environmental Affairs (IPE), a Chinese NGO that specializes in transparency initiatives, has used newly available data to create "incentive structure[s] for win-win outcomes for actors that might otherwise prefer to pollute" (p.17). According to Haddad, transparency initiatives can also lead to "new norms about public disclosure, corporate governance, and consumer responsibility" (p.19). Similarly, Tan (2014) examines two cases—local pollution databases and the Pollution Information Transparency Index (PITI)—and concludes that the OEI has had limited but tangible impacts on data availability. Li (2011) also argues that new environmental disclosure measures have the potential to empower citizens to "seek and use environmental information actively in decision-making and redressing pollution harms" (p.331). Finally, Johnson (2011) suggests that China's OEI regulations provide additional resources for non-state actors as well as a legal basis for their actions.

However, transparency initiatives are often diluted and distorted as they pass through the many layers of China's fragmented bureaucracy. Tan (2014) reports that environmental disclosure is weakest in the most polluted cities and less effective when local governments exhibit low political capacity. Lorentzen, Landry, and Yasuda (2014) reach similar conclusions—they find that environmental transparency measures are less successful in cities dominated by large industrial firms, in particular when the largest firm is in a highly polluting industry. Striking a cautionary note, Tan (2014) warns readers not to equate environmental transparency with policy accountability, because one of these two situations can exist without the other. Therefore, further research is needed to determine how information disclosure actually affects environmental governance outcomes in the Chinese context.

Third-Party Labeling and Corporate Social Responsibility

In addition to providing information for government oversight, environmental transparency initiatives can also help consumers to buy goods that are safer and more sustainably produced. This is a particularly important concern in China, where the scandal over melamine-contaminated milk in 2008 and various other food safety issues have garnered extensive public attention. Third-party regulation put in place following this heightened public concern over food safety plays an increasingly important role in China's governance landscape. Ecolabeling, also known as environmental certification, is one of the most popular forms of third-party regulation (Mol 2014). As of mid-2013, China's National Certification and Accreditation Administration listed 48 certification agencies. The two most widely adopted schemes—the "hazard-free food" and "green food" programs—are certified by independent agencies affiliated with China's Ministry of Agriculture, while most other certification programs are run by NGOs (Zhang et al. 2015).

Other scholars have written about the new movement toward labeling organic products in China, a trend imported from abroad. Although Chinese farmers have traditionally grown organic-labeled products for consumption in Western markets, the growth in China's urban middle class has also led to an increasing appetite for organic food at home (Dendler and Dewick 2016; Yin et al. 2010). However, despite this rapid growth, organic labeling is still in its early stages and accounts for only a minuscule proportion of total domestic food production. Other research shows that most current organic labeling programs fall far short of consumers' expectations. On the one hand, a significant proportion of Chinese consumers are willing to pay more for organic food because of its perceived health, environmental, and taste benefits (Liu et al. 2013; Thøgersen and Zhou 2012). However, several highly publicized incidents of data falsification and corruption (Cheng 2012) have created widespread consumer distrust of Chinese labeling schemes (Yin et al. 2010; Zhang et al. 2015). Dendler and Dewick (2016) show that further institutionalization of organic labeling in China is hindered by limited support from government officials and mainstream retailers as well as a lack of capacity among China's small farmers.

Besides voluntary programs, many Chinese businesses are now required to disclose information on their environmental and social activities to the general public. Since 2001, companies undergoing an initial public offering must report their environmental risks in order to be listed on the Shenzhen Stock Exchange. Shenzhen also became the first Chinese stock exchange to issue guidelines encouraging listed enterprises to produce a corporate social responsibility (CSR) report along with their annual financial report. The Shanghai Stock Exchange later followed Shenzhen's example, with similar requirements and guidelines adopted in 2008 (Noronha et al. 2013). Despite the rapid implementation of these practices, most researchers·believe that, in general, corporate environmental disclosure in China has been rather ineffective. For example, Liu and Anbumozhi (2009) found that 40 percent of the companies in their sample "opened no substantial environmental data to the public" (p.593). Similarly, Kuo et al. (2012) sampled CSR reports from 2010 and concluded that 41 percent of them failed to provide any useful information about the company's CSR activities. Nevertheless, Weber (2014) provides some counterpoint to these arguments, suggesting that the quality and frequency of Chinese environmental disclosure have improved over time because of new government regulations.

Although these studies weigh the effectiveness of environmental disclosure regulations, other scholars have sought to understand the more fundamental drivers of firms' CSR practices. For instance, Liu and Anbumozhi (2009) note that companies in better-off coastal provinces tend to release more information, as do firms with better economic performance. Zeng et al. (2010) write that company size is another important determinant, with larger companies more likely to release information. Meanwhile, Li and Zhang (2010) report that corporate ownership dispersion is likewise positively associated with better CSR practices. Ownership structure plays a similarly important role, and Kuo et al. (2012) note that "environmentally sensitive industries (ESIs) and state-owned enterprises (SOEs) are more dedicated to environmental information disclosure" (p.273). Finally, Peng et al. (2015) argue that firms' decisions about whether to disclose information depend in large part upon the actions of their industrial competitors and peers. However, other researchers report that powerful business and political stakeholders actually have relatively little influence in determining firms' social and environmental disclosure practices (Liu and Anbumozhi 2009; Lu and Abeysekera 2014).

The Rise of Green Finance

In recent years, a great deal of scholarly attention has been focused on the relatively new phenomenon of "green financing" in China (Li and Hu 2014; Zhang et al. 2011). The country's first green finance policy, known as the Green Credit Guidelines, was adopted in 2007 (revised in 2012) and aims to improve environmental quality by restricting loans to companies and for construction projects with poor records of environmental performance. More recently, the State Council released a 2015 report, titled the "Ecological Civilization System Reform Guidelines," which includes specific sections related to green finance. These latest guidelines broaden the scope of green financing and add a wide variety of new market-based tools, such as "green securities," "green bonds," and "green development funds."[13] The guidelines also established compulsory liability insurance mechanisms for firms operating in areas of high environmental risk. Many of these policy tools were initially developed by the Green Finance Task Force, an international partnership supported by the PBOC and the United Nations Environment Programme (UNEP) (Green Finance Task Force 2015).

With respect to implementation, the implementation of China's green finance policies is shared by three government agencies: the MEP, the PBOC, and the China Banking Regulatory Commission (CBRC). The MEP is primarily in charge of establishing an environmental performance information system to be used by all levels of government, while the PBOC oversees the green credit reporting system and sets up other financial information structures. Finally, the CBRC guides and supervises banks so that they will regard environmental

---

13   The "green securities" policy is a national environmental disclosure regulation (approved in 2008) that requires Chinese companies listed on domestic stock exchanges in fourteen highly polluting industries to disclose environmental information. "Green bonds" are fixed financial instruments issued by governments, multinational banks, or corporations to raise money for designated environmental purposes (during the first eight months of 2016, China issued green bonds worth a total of $17.4 billion). Finally, China's adoption of a nationwide "green development fund" was announced by the PBOC on September 1, 2016. The program will set aside central funds for local governments to invest in green infrastructure projects.

compliance as an independent condition for loan approvals (Aizawa and Yang 2010).

Although some observers are enthusiastic about the potential of green finance in China, policy implementation so far has been lackluster. Zhang et al. (2011) identify unclear implementation standards and a lack of reliable information as the major barriers to advancing green credit schemes in China, which other scholars attribute to the lack of incentives and weak supervision by financial institutions (Li and Hu 2014). Thus, if the government truly wishes to harness the power of financial markets to stimulate environmental protection, many additional policy fixes are required—including, but not limited to, clearer laws, more transparent information, and better coordination among agencies.

Indeed, similar sentiments can be applied to all of China's market-based environmental governance policies. Whether with respect to SO_2 trading schemes, carbon markets, PES programs, or information disclosure initiatives, the benefits of market efficiency must be weighed against the need for government supervision and control. In China, market-based environmental protection measures are still determined more often by administrative targets than by supply and demand (Goron and Cassica 2017). Although this by no means renders these policies ineffective, researchers must be careful to account for China's unique state-market relationship when they attempt to understand the meaning of subsequent environmental outcomes.

5 Conclusion

In March 2014, Premier Li Keqiang declared a "war on pollution" and vowed that China would fight pollution as it did poverty (Reuters 2014). The following December, China, together with the United States and 193 other countries, adopted the Paris climate agreement (under the United Nations Framework Convention on Climate Change) and kick-started a global effort to combat climate change. Then, at the annual G20 summit, held in the ancient, scenic lakeside city of Hangzhou in September 2016, President Xi Jinping and President Barack Obama formally expressed their commitment to the Paris agreement (Landler and Perlez 2016). However, just two months later, hopes for greater US-China collaboration on climate change were dashed when Donald Trump was elected to succeed Obama as president of the United States. After President Trump announced the US withdrawal from the Paris agreement, many observers speculated that China could end up taking the lead in the global fight against climate change (Worland 2017). Xie Zhenghua, a former NEPA and NDRC official and China's negotiator at several UN climate change conferences, declared after Trump's election that China would "act as facilitator to boost climate negotiations on implementing the Paris Agreement, regardless of the stance the United States takes" (Geall 2016). Indeed, experts report that China is already well on its way to achieving its coal peak by 2030; its coal consumption decreased by 2.9 percent in 2014 and 3.6 percent in 2015 (Lu and Ye 2016).

With these recent developments in mind, scholars continue to grapple with important research questions in the field of Chinese environmental governance. For instance, what explains institutional change in the context of Chinese environmental governance? Which actors are responsible for bringing about these changes, and what factors determine their incentives and motivate their behavior? Finally, which institutions are the most effective at producing positive changes?

In many ways, China's environmental governance system is a study in contradictions. The country's environmental policy regime is a complex, fragmented, and cumbersome apparatus yet is also surprisingly flexible in its operation. Environmental management in China is highly decentralized yet also exhibits some consistent patterns of behavior. With respect to local incentives, the same system that brought about four decades of economic development has proven to be ineffective—if not counterproductive—in controlling pollution. Finally, local EPBs have gained increased capacity to enforce policy, yet they often still appear powerless in the face of political, economic, and social forces beyond their control. In such a complex and variegated system, bargaining and negotiation are often the keys to successful environmental policy-making and implementation. With this in mind, scholars have observed patterns of bargaining among actors within the state, between state and societal actors, and between the state and the market. Future work should delve further into these interrelationships and attempt to uncover the incentives and strategies of various actors.

When analyzing Chinese environmental governance, perhaps the most important truism is that the environmental state is ever-evolving. As we write this article, new institutional innovations are continuously being proposed, tested, and implemented. These innovations range from bottom-up local initiatives to top-down experiments, and most of their effects have yet to be seen. It is therefore incumbent upon students of China's environmental state to follow current and future institutional developments within the country closely and observe their effects on the behavior of key stakeholders.

References

Aizawa, Motoko, and Chaofei Yang. 2010. "Green Credit, Green Stimulus, Green Revolution? China's Mobilization of Banks for Environmental Cleanup." *Journal of Environment & Development* 19: 119–44.

Balme, Richard, and Renwu Tang. 2014. "Environmental Governance in the People's Republic of China: The Political Economy of Growth, Collective Action, and Policy Developments—Introductory Perspectives." *Asian Pacific Journal of Public Administration* 36: 167–72.

Barboza, David, and Sharon LaFraniere. 2012. "'Princelings' in China Use Family Ties to Gain Riches." *New York Times*, May 17. Accessed November 11, 2016. http://www.nytimes.com/2012/05/18/world/asia/china-princelings-using-family-ties-to-gain-riches.html.

Baum, Richard, and Alexei Shevchenko. 1999. "The State of the State." In *The Paradox of China's Post-Mao Reforms*, eds. Merle Goldman and Roderick MacFarquhar, pp. 333–60. Cambridge: Harvard University Press.

Beeson, Mark. 2010. "The Coming of Environmental Authoritarianism." *Environmental Politics* 19: 276–94.

Bennett, Michael T. 2008. "China's Sloping Land Conversion Program: Institutional Innovation or Business as Usual?" *Ecological Economics* 65: 699–711.

Brauer, Michael. 2016. "The Global Burden of Disease from Air Pollution." Paper presented at the annual meeting of the AAAS. Washington, DC: February 11–15.

Brettell, Anna. 2003. "The Politics of Public Participation and the Emergence of Environmental Proto-Movements in China." Ph.D. diss., University of Maryland.

Carter, Neil, and Arthur P.J. Mol. 2013. *Environmental Governance in China*. London: Routledge.

Carter, Neil, and Arthur P.J. Mol. 2006. "China and the Environment: Domestic and Transnational Dynamics of a Future Hegemon." *Environmental Politics* 15: 330–44.

CDM Pipeline. 2016. http://www.cdmpipeline.org/cdm-projects-region.htm. Accessed December 31.

Chen, Jie, 2010. "Transnational Environmental Movement: Impacts on the Green Civil Society in China." *Journal of Contemporary China* 19: 503–23.

Chen, Xiaodong, Nils Peterson, Vanessa Hull, Chuntian Lu, Dayong Hong, and Jianguo Liu. 2013. "How Perceived Exposure to Environmental Harm Influences Environmental Behavior in Urban China." *Ambio* 42: 52–60.

Chen, Ye, Hongbin Li, and Li-An Zhou. 2005. "Relative Performance Evaluation and the Turnover of Provincial Leaders in China." *Economics Letters* 88: 421–25.

Cheng, Hongming. 2012. "Cheap Capitalism: A Sociological Study of Food Crime in China." *British Journal of Criminology* 52: 254–73.

CNNIC. 2016. "Statistical Report on Internet Development in China." Accessed December 12, 2016. https://cnnic.com.cn/IDR/ReportDownloads/201611/P020161114 573409551742.pdf.

Cross, Remy, and David Snow. 2010. "Social Movements." In *The Wiley-Blackwell Companion to Sociology*, ed. George Ritzer, pp. 522–44. Hoboken, NJ: John Wiley & Sons.

Cunningham, Edward. 2015. "The State and the Firm: Chinese Energy Governance in Context." CEGI Working Paper. Boston, MA.

Dasgupta, Susmita, and David Wheeler. 1997. "Citizen Complaints as Environmental Indicators: Evidence from China." World Bank Policy Research Working Paper. Washington, DC.

Delang, Claudio O., and Zhen Yuan. 2015. *China's Grain for Green Program: A Review of the Largest Ecological Restoration and Rural Development Program in the World.* Cham, Switzerland: Springer International.

Dendler, Leonie, and Paul Dewick. 2016. "Institutionalising the Organic Labelling Scheme in China: A Legitimacy Perspective." *Journal of Cleaner Production* 134: 239–50.

Deng, Lei, Guo-bin Liu, and Zhou-ping Shangguan. 2014. "Land-Use Conversion and Changing Soil Carbon Stocks in China's 'Grain-for-Green' Program: A Synthesis." *Global Change Biology* 20: 3544–56.

Deng, Yanhua, and Guobin Yang. 2013. "Pollution and Protest in China: Environmental Mobilization in Context." *China Quarterly*, no. 214: 321–36.

Ding, Iza. 2016. "Invisible Sky, Visible State: Environmental Governance and Political Support in China." Ph.D. diss., Harvard University.

Ding, Iza. 2017. "Performative Governance." Working Paper. University of Pittsburgh.

Dong, Yanli, Masanobu Ishikawa, Xianbing Liu, and Shigeyuki Hamori. 2011. "The Determinants of Citizen Complaints on Environmental Pollution: An Empirical Study from China." *Journal of Cleaner Production* 19: 1306–14.

Downs, Erica, and Michal Meidan. 2011. "Business and Politics in China." *China Security* 19: 3–21.

Dunlap, Riley E., Kent D. Van Liere, Angela G. Mertig, and Robert Emmet Jones. 2000. "New Trends in Measuring Environmental Attitudes: Measuring Endorsement of the New Ecological Paradigm: A Revised NEP Scale." *Journal of Social Issues* 56: 425–42.

Eaton, Sarah, and Genia Kostka. 2014. "Authoritarian Environmentalism Undermined? Local Leaders' Time Horizons and Environmental Policy Implementation in China." *China Quarterly*, no. 218: 359–80.

Economy, Elizabeth. 2004. *The River Runs Black: The Environmental Challenge to China's Future.* Ithaca: Cornell University Press.

Economy, Elizabeth. 2006. "Environmental Governance: The Emerging Economic Dimension." *Environmental Politics* 15: 171–89.

Economy, Elizabeth. 2013. "China's New Governing Style: Crisis Management." *The Atlantic*, May 20. Accessed December 29, 2016. http://www.theatlantic.com/china/archive/2013/05/chinas-new-governing-style-crisis-management/276034/.

Economy, Elizabeth. 2014. "Environmental Governance in China: State Control to Crisis Management." *Daedalus* 143: 184–97.

Edin, Maria. 2003. "State Capacity and Local Agent Control in China: CCP Cadre Management from a Township Perspective." *China Quarterly*, no. 173: 35–52.

Edmonds, Richard. 1994. *Patterns of China's Lost Harmony: A Survey of the Country's Environmental Degradation and Protection*. London: Routledge.

Elvin, Mark. 1998. "The Environmental Legacy of Imperial China." *China Quarterly*, no. 156: 733–56.

Elvin, Mark. 2008. *The Retreat of the Elephants: An Environmental History of China*. New Haven: Yale University Press.

Emissions Database for Global Atmospheric Research. 2016. "CO_2 Emissions Time Series 1990–2015 per Region/Country." Accessed December 16, 2016. http://edgar.jrc.ec.europa.eu/overview.php?v=CO2ts1990-2015.

Fang, Xiuqi, Yun Su, Jun Yin, and Jingchao Teng. 2015. "Transmission of Climate Change Impacts from Temperature Change to Grain Harvests, Famines and Peasant Uprisings in the Historical China." *Science China Earth Sciences* 58: 1427–39.

Fedorenko, Irina, and Yixian Sun. 2016. "Microblogging-based Civic Participation on Environment in China: A Case Study of the $PM_{2.5}$ Campaign." *VOLUNTAS: International Journal of Voluntary and Nonprofit Organizations* 27: 2077–2105.

Feng, Lin, and Jianying Xu. 2015. "Farmers' Willingness to Participate in the Next-Stage Grain-for-Green Project in the Three Gorges Reservoir Area, China." *Environmental Management* 56: 505–18.

Feng, Lu, and Wenjie Liao. 2016. "Legislation, Plans, and Policies for Prevention and Control of Air Pollution in China: Achievements, Challenges, and Improvements." *Journal of Cleaner Production* 112: 1549–58.

Fishkin, James S., Baogang He, Robert C. Luskin, and Alice Siu. 2010. "Deliberative Democracy in an Unlikely Place: Deliberative Polling in China." *British Journal of Political Science* 40, no. 2: 435–48.

Frolic, Michael. 1997. "State-Led Civil Society." In *Civil Society in China*, ed. Timothy Brook and Michael Frolic, pp. 46–67. Armonk, NY: M.E. Sharpe.

Gallagher, Mary. 2007. "Hope for Protection and Hopeless Choices." In *Grassroots Political Reform in Contemporary China*, ed. Merle Goldman and Elizabeth Perry, pp. 196–227. Cambridge: Harvard University Press.

Gaventa, John, and Rosemary McGee. 2013. "The Impact of Transparency and Accountability Initiatives." *Development Policy Review* 31: 3–28.

Geall, Sam. 2016. "Comments: Will China Take the Lead on Climate Change?" China File, November 21. Accessed December 29, 2016. https://www.chinafile.com/conversation/will-china-take-lead-climate-change.

Gilley, Bruce. 2012. "Authoritarian Environmentalism and China's Response to Climate Change." *Environmental Politics* 21: 287–307.

Goron, Coraline, and Cyril Cassisa. 2016. "Regulatory Institutions and Market-based Climate Policy in China." *Global Environmental Politics*: 99–120.

Grano, Simona, and Yuheng Zhang. 2016. "New Channels for Popular Participation in China: The Case of an Environmental Protection Movement in Nanjing." *China Information* 30: 129–42.

Green Finance Task Force. 2015. *Establishing China's Green Financial System.* Accessed December 29, 2016. https://www.cbd.int/financial/privatesector/china-Green%20Task%20Force%20Report.pdf.

Gu, Edward. 2011. "Towards a Corporatist Development of Civil Society in China: Enabling State and Mutual Empowerment between State and Society." In *State and Civil Society: The Chinese Perspective*, ed. Zhenglai Deng, pp. 385–405. Singapore: World Scientific Publishing.

Guan, Dabo, Zhu Liu, Yong Geng, Sören Lindner, and Klaus Hubacek. 2012. "The Gigatonne Gap in China's Carbon Dioxide Inventories." *Nature Climate Change* 2: 672–75.

Gupta, Aarti. 2010. "Transparency in Global Environmental Governance: A Coming of Age?" *Global Environmental Politics* 10: 1–9.

Haddad, Mary Alice. 2015. "Increasing Environmental Performance in a Context of Low Governmental Enforcement: Evidence from China." *Journal of Environment & Development* 24: 3–25.

Harris, Paul G. 2006. "Environmental Perspectives and Behavior in China Synopsis and Bibliography." *Environment and Behavior* 38: 5–21.

Haufler, Virginia. 2010. "Disclosure as Governance: The Extractive Industries Transparency Initiative and Resource Management in the Developing World." *Global Environmental Politics* 10: 53–73.

He, Baogang, and Mark E. Warren. 2011. "Authoritarian Deliberation: The Deliberative Turn in Chinese Political Development." *Perspectives on Politics* 9, no. 2: 269–89.

He, Baogang, and Stig Thogersen. 2010. "Giving the People a Voice? Experiments with Consultative Authoritarian Institutions in China." *Journal of Contemporary China* 19: 675–692.

Heilmann, Sebastian, and Elizabeth J. Perry. 2011. "Embracing Uncertainty: Guerrilla Policy Style and Adaptive Governance in China." In *Mao's Invisible Hand: The Political Foundation of Adaptive Governance in China*, eds. Sebastian Heilmann and Elizabeth Perry, pp. 1–29. Cambridge: Harvard University Press.

Hildebrandt, Timothy. 2013. *Social Organizations and the Authoritarian State in China*. New York: Cambridge University Press.

Ho, Peter. 2001. "Greening without Conflict? Environmentalism, NGOs and Civil Society in China." *Development and Change* 32: 893–921.

Ho, Peter. 2007. "Self-Imposed Censorship and De-Politicized Politics in China." In *China's Embedded Activism: Opportunities and Constraints of a Social Movement*, eds. Peter Ho and Richard Edmonds, pp. 20–44. London: Routledge.

Ho, Peter, and Richard Edmonds. 2007. *China's Embedded Activism: Opportunities and Constraints of a Social Movement*. London: Routledge.

Hoffman, Samantha, and Jonathan Sullivan. 2015. "Environmental Protests Expose Weakness in China's Leadership." *Forbes Asia*, June 22. Accessed December 29, 2016. http://www.forbes.com/sites/forbesasia/2015/06/22/environmental-protests-expose-weakness-in-chinas-leadership/#32401d832f09/.

Holdaway, Jennifer. 2013. "Environment and Health Research in China: The State of the Field." *China Quarterly*, no. 214: 255–82.

Horizon Key. 2015. "China Public Evaluation of Government and Government Public Services Research Report." Beijing Horizon Key Information Consulting, LLC. Report Prepared for the Harvard Ash Center for Democratic Governance and Innovation. February 1.

Hsu, Angel, Alex de Sherbinin, and Han Shi. 2012. "Seeking Truth from Facts: The Challenge of Environmental Indicator Development in China." *Environmental Development* 3: 39–51.

Hsu, Jennifer. 2014. "Chinese Non-Governmental Organizations and Civil Society: A Review of the Literature." *Geography Compass* 8: 98–110.

Hsu, Jennifer, and Reza Hasmath. 2014. "The Local Corporatist State and NGO Relations in China." *Journal of Contemporary China* 23: 516–34.

Huang, Ganlin. 2015. "$PM_{2.5}$ Opened a Door to Public Participation Addressing Environmental Challenges in China." *Environmental Pollution* 197: 313–15.

Huang, Lei, Jie Ban, Kai Sun, Yuting Han, Zengwei Yuan, and Jun Bi. 2013. "The Influence of Public Perception on Risk Acceptance of the Chemical Industry and the Assistance for Risk Communication." *Safety Science* 51: 232–40.

Huang, Yasheng. 1996. *Inflation and Investment Controls in China: The Political Economy of Central-Local Relations during the Reform Era*. New York: Cambridge University Press.

Jahiel, Abigail R. 1998. "The Organization of Environmental Protection in China." *China Quarterly*, no. 156: 757–87.

Jian, Lu, and Chris King-Chi Chan. 2016. "Collective Identity, Framing and Mobilisation of Environmental Protests in Urban China: A Case Study of Qidong's Protest." *China: An International Journal* 14: 102–22.

Jiang, Jingjing, Dejun Xie, Bin Ye, Bo Shen, and Zhanming Chen. 2016. "Research on China's Cap-and-Trade Carbon Emission Trading Scheme: Overview and Outlook." *Applied Energy* 178: 902–17.

Jiang, Min. 2016. "The Co-Evolution of the Internet, (Un)Civil Society, and Authoritarianism in China." In *The Internet, Social Media, and a Changing China*, ed. Avery Goldstein, and Guobin Yang, pp. 22–48. Philadelphia: University of Pennsylvania Press.

Jin, Zining. 2015. "Environmental Impact Assessment Law in China's Courts: A Study of 107 Judicial Decisions." *Environmental Impact Assessment Review* 55: 35–44.

Jing, Jun. 2003. "Environmental Protests in Rural China." In *Chinese Society: Change, Conflict and Resistance*, ed. Elizabeth Perry and Mark Selden, pp. 208–226. London: Routledge Curzon.

Johnson, Thomas R. 2010. "Environmentalism and NIMBYism in China: Promoting a Rules-Based Approach to Public Participation." *Environmental Politics* 19: 430–48.

Johnson, Thomas R. 2011. "Environmental Information Disclosure in China: Policy Developments and NGO Responses." *Policy & Politics* 39: 399–416.

Jotzo, Frank, and Andreas Löschel. 2014. "Emissions Trading in China: Emerging Experiences and International Lessons." *Energy Policy* 75: 3–8.

Kay, Samuel, Bo Zhao, and Daniel Sui. 2015. "Can Social Media Clear the Air? A Case Study of the Air Pollution Problem in Chinese Cities." *Professional Geographer* 67: 351–63.

Kennedy, John. 2012. "Environmental Protests in China on Dramatic Rise, Expert Says." *South China Morning Post*, October 29. Accessed December 29, 2016. http://www.scmp.com/news/china/article/1072407/environmental-protests-china-rise-expert-says/.

Kennedy, John James. 2007. "The Implementation of Village Elections and Tax-for-Fee Reform in Rural Northwest China." *Harvard Contemporary China Series* 14: 48–67.

King, Gary, Jennifer Pan, and Margaret E. Roberts. 2013. "How Censorship in China Allows Government Criticism but Silences Collective Expression." *American Political Science Review* 107: 326–43.

Kostka, Genia. 2013. "Environmental Protection Bureau Leadership at the Provincial Level in China: Examining Diverging Career Backgrounds and Appointment Patterns." *Journal of Environmental Policy & Planning* 15: 41–63.

Kostka, Genia. 2016. "Command without Control: The Case of China's Environmental Target System." *Regulation & Governance* 10: 58–74.

Kostka, Genia, and Arthur P. Mol. 2013. "Implementation and Participation in China's Local Environmental Politics: Challenges and Innovations." *Journal of Environmental Policy & Planning* 15: 3–16.

Kung, James Kai-sing, and Shuo Chen. 2011. "The Tragedy of the Nomenklatura: Career Incentives and Political Radicalism during China's Great Leap Famine." *American Political Science Review* 105: 27–45.

Kuo, Lopin, Chin-Chen Yeh, and Hui-Cheng Yu. 2012. "Disclosure of Corporate Social Responsibility and Environmental Management: Evidence from China." *Corporate Social Responsibility and Environmental Management* 19: 273–87.

Landler, Mark, and Jane Perlez. 2016. "Rare Harmony as China and U.S. Commit to Climate Deal." *New York Times*, September 3. Accessed December 29, 2016. http://www.nytimes.com/2016/09/04/world/asia/obama-xi-jinping-china-climate-accord.html.

Larson, Christina. 2015. "China Hails, Then Bans a Documentary." Bloomberg News, March 12. Accessed December 29, 2016. https://www.bloomberg.com/news/articles/2015-03-12/china-hails-then-bans-antipollution-film-under-the-dome/.

Leene, Elora, Emy Marier, Wes Meives, David Hahn, and Helue Vazquez Valverde. 2014. "Examining Adult Public Opinion on Climate Change in the United States and China." University of Wisconsin Eau-Claire Economics Department report.

Li, Aitong. 2016. "Forests or Grasslands: Conflicts over Environmental Conservation in Arid and Semiarid Regions of China." *Regional Environmental Change* 16: 1535–39.

Li, Wanxin. 2011. "Self-Motivated versus Forced Disclosure of Environmental Information in China: A Comparative Case Study of the Pilot Disclosure Programmes." *China Quarterly*, no. 206: 331–51.

Li, Wanxin, Jieyan Liu, and Duoduo Li. 2012. "Getting Their Voices Heard: Three Cases of Public Participation in Environmental Protection in China." *Journal of Environmental Management* 98: 65–72.

Li, Wei, and Mengze Hu. 2014. "An Overview of the Environmental Finance Policies in China: Retrofitting an Integrated Mechanism for Environmental Management." *Frontiers of Environmental Science & Engineering* 8: 316–28.

Li, Wenjing, and Ran Zhang. 2010. "Corporate Social Responsibility, Ownership Structure, and Political Interference: Evidence from China." *Journal of Business Ethics* 96: 631–45.

Li, Yanwei, Joop Koppenjan, and Stefan Verweij. 2016. "Governing Environmental Conflicts in China: Under What Conditions Do Local Governments Compromise?" *Public Administration* 94: 806–22.

Liang, Yicheng, Shuzhuo Li, Marcus W. Feldman, and Gretchen C. Daily. 2012. "Does Household Composition Matter? The Impact of the Grain for Green Program on Rural Livelihoods in China." *Ecological Economics* 75: 152–60.

Liao, Janet. 2014. "The Chinese Government and the National Oil Companies (NOCs): Who Is the Principal?" *Asian Pacific Business Review* 21: 44–59.

Lieberthal, Kenneth, and David M. Lampton. 1992. *Bureaucracy, Politics, and Decision Making in Post-Mao China*. Berkeley: University of California Press.

Lieberthal, Kenneth, and Michel Oksenberg. 1998. *Policy Making in China: Leaders, Structures, and Processes*. Princeton: Princeton University Press.

Liebman, Benjamin. 2007. "China's Courts: Restricted Reform." *China Quarterly*, no. 191: 620–38.

Liebman, Benjamin. 2011. "A Populist Threat to China's Courts?" In *Chinese Justice: Civil Dispute Resolution in Contemporary China*, eds. Margaret Woo and Mary Gallagher, pp. 269–314. New York: Cambridge University Press.

Liebold, James. 2011. "Blogging Alone: China, the Internet, and the Democratic Delusion?" *Journal of Asian Studies* 70: 1023–41.

Lin, Li, and Tian He. 2014. *Zhonguo fazhi fazhan baogao* 中国法制发展报告 [Annual Report on China's Rule of Law]. Beijing: Chinese Academy of Social Sciences Press.

Lin, Ying, and Shunbo Yao. 2014. "Impact of the Sloping Land Conversion Program on Rural Household Income: An Integrated Estimation." *Land Use Policy* 40: 56–63.

Liu, Can, Jinzhi Lu, and Runsheng Yin. 2010. "An Estimation of the Effects of China's Priority Forestry Programs on Farmers' Income." *Environmental Management* 45: 526–40.

Liu, Huaxing. 2015. "Why Is There Less Public Trust in Local Government Than in Central Government in China?" Ph.D. diss., University of Birmingham.

Liu, Jianguo, Shuxin Li, Zhiyun Ouyang, Christine Tam, and Xiaodong Chen. 2008. "Ecological and Socioeconomic Effects of China's Policies for Ecosystem Services." *Proceedings of the National Academy of Sciences* 105: 9477–82.

Liu, John Chung-En, and Anthony A. Leiserowitz. 2009. "From Red to Green?" *Environment: Science and Policy for Sustainable Development* 51: 32–45.

Liu, Nicole Ning, Carlos Wing-Hung Lo, Xueyong Zhan, and Wei Wang. 2014. "Campaign-Style Enforcement and Regulatory Compliance." *Public Administration Review* 75: 85–95.

Liu, Rongduo, Zuzanna Pieniak, and Wim Verbeke. 2013. "Consumers' Attitudes and Behaviour towards Safe Food in China: A Review." *Food Control* 33: 93–104.

Liu, Xianbing, and Venkatachalam Anbumozhi. 2009. "Determinant Factors of Corporate Environmental Information Disclosure: An Empirical Study of Chinese Listed Companies." *Journal of Cleaner Production* 17: 593–600.

Liu, Xianbing, Qinqin Yu, Tetsuro Fujitsuka, Beibei Liu, Jun Bi, and Tomohiro Shishime. 2010. "Functional Mechanisms of Mandatory Corporate Environmental Disclosure: An Empirical Study in China." *Journal of Cleaner Production* 18: 823–32.

Liu, Xinsheng, and Ren Mu. 2016. "Public Environmental Concern in China: Determinants and Variations." *Global Environmental Change* 37: 116–27.

Liu, Zhu, Dabo Guan, Wei Wei, Steven J. Davis, Philippe Ciais, Jin Bai, Shushi Peng, Qiang Zhang, Klaus Hubacek, Gregg Marland, Robert J. Andres, Douglas Crawford-Brown, Jintai Lin, Hongyan Zhao, Chaopeng Hong, Thomas A. Boden, Kuishuang Feng, Glen P. Peters, Fengming Xi, Junguo Liu, Yuan Li, Yu Zhao, Ning Zeng, and Kebin He. 2015. "Reduced Carbon Emission Estimates from Fossil Fuel Combustion and Cement Production in China." *Nature* 524: 335–38.

Lo, Alex Y. 2015. "National Development and Carbon Trading: The Symbolism of Chinese Climate Capitalism." *Eurasian Geography and Economics* 56: 111–26.

Lo, Alex Y. 2016. "Challenges to the Development of Carbon Markets in China." *Climate Policy* 16: 109–24.

Lo, Alex Y., and Michael Howes. 2013. "Powered by the State or Finance? The Organization of China's Carbon Markets." *Eurasian Geography and Economics* 54: 386–408.

Lo, Carlos W.H., and Shui-Yan Tang. 1994. "Institutional Contexts of Environmental Management: Water Pollution Control in Guangzhou, China." *Public Administration and Development* 14: 53–64.

Lo, Carlos W.H., and Shui-Yan Tang. 2006. "Institutional Reform, Economic Changes, and Local Environmental Management in China: The Case of Guangdong Province." *Environmental Politics* 15: 190–210.

Lora-Wainwright, Anna. 2013. "Dying for Development: Pollution, Illness and the Limits of Citizens' Agency in China." *China Quarterly*, no. 214: 243–54.

Lora-Wainwright, Anna, Yiyun Zhang, Yunmei Wu, and Benjamin van Rooij. 2012. "Learning to Live with Pollution: The Making of Environmental Subjects in a Chinese Industrialized Village." *China Journal* 68: 106–24.

Lorentzen, Peter, Pierre Landry, and John Yasuda. 2014. "Undermining Authoritarian Innovation: The Power of China's Industrial Giants." *Journal of Politics* 76: 182–94.

Lu, Jiaqi, and Qi Ye. 2016. "The End of Coal-Fired Growth in China." Brookings Institution, August 4. Accessed December 29, 2016. https://www.brookings.edu/blog/up-front/2016/08/04/the-end-of-coal-fired-growth-in-china/.

Lu, Yihe, Bojie Fu, Xiaoming Feng, Yuan Zeng, Yu Liu, Ruiying Chang, Ge Sun, and Bingfang Wu. 2012. "A Policy-Driven Large Scale Ecological Restoration: Quantifying Ecosystem Services Changes in the Loess Plateau of China." *PloS One* 7: 1–10.

Lu, Yingjun, and Indra Abeysekera. 2014. "Stakeholders' Power, Corporate Characteristics, and Social and Environmental Disclosure: Evidence from China." *Journal of Cleaner Production* 64: 426–36.

Ma, Xiaoying, and Leonard Ortolano. 2000. *Environmental Regulation in China: Institutions, Enforcement, and Compliance*. Lanham, MD: Rowman & Littlefield.

MacKinnon, Rebecca. 2012. "China's 'Networked Authoritarianism.'" In *Liberation Technology*, eds. Larry Diamond and Marc Plattner, pp. 78–94. Baltimore: Johns Hopkins University Press.

Manion, Melanie. 1985. "The Cadre Management System, Post-Mao: The Appointment, Promotion, Transfer and Removal of Party and State Leaders." *China Quarterly*, no. 102: 203–33.

Marks, Robert B. 2011. *China: its Environment and History*. Lanham, MD: Rowman & Littlefield.

Martens, Susan. 2006. "Public Participation with Chinese Characteristics: Citizen Consumers in China's Environmental Management." *Environmental Politics* 15: 211–30.

Mason, Michael. 2008. "Transparency for Whom? Information Disclosure and Power in Global Environmental Governance." *Global Environmental Politics* 8: 8–13.

Merchant, Carolyn. 1980. *The Death of Nature: Women, Ecology, and Scientific Revolution.* New York: Harper & Row.

Mertha, Andrew. 2008. *China's Water Warriors: Citizen Action and Policy Change.* Ithaca: Cornell University Press.

Mertha, Andrew. 2009. "'Fragmented Authoritarianism 2.0': Political Pluralization in the Chinese Policy Process." *China Quarterly*, no. 200: 995–1012.

Millman, Alexander, Deliang Tang, and Frederica P. Perera. 2008. "Air Pollution Threatens the Health of Children in China." *Pediatrics* 122: 620–628.

Mol, Arthur P.J., and Neil T. Carter. 2006. "China's Environmental Governance in Transition." *Environmental Politics* 15: 149–70.

Mol, Arthur P.J., Guizhen He, and Lei Zhang. 2011. "Information Disclosure in Environmental Risk Management: Developments in China." *Journal of Current Chinese Affairs* 40: 163–92.

Munnings, Clayton, Richard D. Morgenstern, Zhongmin Wang, and Xu Liu. 2016. "Assessing the Design of Three Carbon Trading Pilot Programs in China." *Energy Policy* 96: 688–99.

Nathan, Andrew J. 2003. "Authoritarian Resilience." *Journal of Democracy* 14: 6–17.

Noronha, Carlos, Si Tou, M.I. Cynthia, and Jenny J. Guan. 2013. "Corporate Social Responsibility Reporting in China: An Overview and Comparison with Major Trends." *Corporate Social Responsibility and Environmental Management* 20: 29–42.

O'Brien, Kevin, and Lianjiang Li. 2006. *Rightful Resistance in Rural China.* Cambridge: Cambridge University Press.

Page, Jeremy, Bryan Spegele, and Wayne Ma. 2013. "Powerful Oil Clique at Center of Chinese Probes: Some See Party Struggle Behind Investigations Targeting Central Pillar of State Sector." *Wall Street Journal*, September 5. Accessed December 29, 2016. http://www.wsj.com/articles/SB10001424127887323623304579056883184524514/.

Pei, Minxin. 2006. *China's Trapped Transition: The Limits of Developmental Autocracy.* Cambridge: Harvard University Press.

Peng, Juan, Jianfei Sun, and Rui Luo. 2015. "Corporate Voluntary Carbon Information Disclosure: Evidence from China's Listed Companies." *World Economy* 38: 91–109.

Piovani, Chiara. 2015. "The 'Greening' of China: Progress, Limitations, and Contradictions." Working Paper. University of Denver, Department of Economics: 1–43.

Qi, Ye. 2008. *Zhongguo huanjing jianguan tizhi yanjiu* 中国环境监管体制研究 [Research on China's Environmental Governance]. Shanghai: Shanghai sanlian Press.

Qu, Geping and Li Jinchang. 1994. *Population and the Environment in China*. Trans. Kiang Batching and Go Ran. CO: Boulder: Lynne Rienner.

Ran, Ran. 2013. "Perverse Incentive Structure and Policy Implementation Gap in China's Local Environmental Politics." *Journal of Environmental Policy & Planning* 15: 17–39.

Reilly, James. 2013. *Strong Society, Smart State: The Rise of Public Opinion in China's Japan Policy*. New York: Columbia University Press.

Reuters. 2014. "China to 'Declare War' on Pollution, Premier Says." March 4. Accessed December 29, 2016. http://www.reuters.com/article/us-china-parliament-pollution-idUSBREA2405W20140305/.

Rohde, Robert, and Richard Muller. 2015. "Air Pollution in China: Mapping of Concentrations and Sources." *PloS One* 10, no. 8: 1–20.

Ross, Lester. 1992. "The Politics of Environmental Policy in the People's Republic of China." *Policy Studies Journal* 20: 628–42.

Saich, Tony. 2015. *Governance and Politics of China*. New York: Palgrave Macmillan.

Schwartz, Jonathan. 2004. "Environmental NGOs in China: Roles and Limits." *Pacific Affairs* 77: 28–49.

Shapiro, Judith. 2001. *Mao's War Against Nature: Politics and the Environment in Revolutionary China*. Cambridge: Cambridge University Press.

Shapiro, Judith. 2012. *China's Environmental Challenges*. Cambridge: Polity Press.

Shieh, Shawn, and Deng Guosheng. 2011. "An Emerging Civil Society: The Impact of the 2008 Sichuan Earthquake on Grassroots Associations in China." *China Journal* 65: 181–94.

Shih, Victor, Christopher Adolph, and Mingxing Liu. 2012. "Getting Ahead in the Communist Party: Explaining the Advancement of Central Committee Members in China." *American Political Science Review* 106: 166–87.

Shin, Sangbum. 2013. "China's Failure of Policy Innovation: The Case of Sulphur Dioxide Emission Trading." *Environmental Politics* 22: 918–34.

Silva, Jérôme, Femke de Keulenaer, and Nick Johnstone. 2012. "Environmental Quality and Life Satisfaction: Evidence Based on Micro-Data." *OECD Publishing* 44: 1–40.

Sinkule, Barbara J., and Leonard Ortolano. 1995. *Implementing Environmental Policy in China*. Westport, CT: Praeger.

Smil, Vaclav. 1984. *The Bad Earth. Environmental Degradation in China*. Armonk, NY: M.E. Sharpe.

Smyth, Russell, Ingrid Nielsen, Qingguo Zhai, Tiemin Liu, Yin Liu, Chunyong Tang, Zhihong Wang, Zuxiang Wang, and Juyong Zhang. 2011. "A Study of the Impact of Environmental Surroundings on Personal Well-Being in Urban China Using a Multi-Item Well-Being Indicator." *Population and Environment* 32: 353–75.

Song, Xinzhang, Changhui Peng, Guomo Zhou, Hong Jiang, and Weifeng Wang. 2014. "Chinese Grain for Green Program Led to Highly Increased Soil Organic Carbon Levels: A Meta-Analysis." *Scientific Reports* 4: 1–7.

Spires, Anthony. 2011. "Contingent Symbiosis and Civil Society in an Authoritarian State: Understanding the Survival of China's Grassroots NGOs." *American Journal of Sociology* 117: 1–45.

State Council Leading Group on Environmental Protection. 1978. *Huanjing baohu gongzuo huibao yaodian* 环境保护工作汇报要点 [Key Points on Environmental Protection Work]. Accessed December 29, 2016. http://www.reformdata.org/content/20131203/23673.html.

Steinhardt, Christoph, and Fengshi Wu. 2016. "In the Name of the Public: Environmental Protest and the Changing Landscape of Popular Contention in China." *China Journal* 75: 61–82.

Stern, Rachel. 2013. *Environmental Litigation in China: A Study in Political Ambivalence.* New York: Cambridge University Press.

Stern, Rachel. 2014. "The Political Logic of China's New Environmental Courts." *China Journal* 72: 53–74.

Stockmann, Daniela. 2012. *Media Commercialization and Authoritarian Rule in China.* New York: Cambridge University Press.

Su, Yang, and Xin He. 2010. "Street as Courtroom: State Accommodation of Labor Protest in South China." *Law and Society Review* 44: 157–84.

Tan, Yeling. 2014. "Transparency without Democracy: The Unexpected Effects of China's Environmental Disclosure Policy." *Governance* 27: 37–62.

Tang, Zhi, and Jintong Tang. 2013. "Can the Media Discipline Chinese Firms' Pollution Behaviors? The Mediating Effects of the Public and Government." *Journal of Management* 1: 1–23.

Tao, Jiaqing, and Xueyong Ai. 1973. "Huanjing baohu gongzuo yao zhuajin 环境保护工作要抓紧 [Environmental Protection Work Must Be Speeded Up]." *People's Daily*, October 15.

Tao, Julia, and Daphne Ngar-yin Mah. 2009. "Between Market and State: Dilemmas of Environmental Governance in China's Sulphur Dioxide Emission Trading System." *Environment and Planning C: Government and Policy* 27: 175–88.

Teets, Jessica. 2014. *Civil Society under Authoritarianism: The China Model.* New York: Cambridge University Press.

Teng, Fei, Xin Wang, and L.V. Zhiqiang. 2014. "Introducing the Emissions Trading System to China's Electricity Sector: Challenges and Opportunities." *Energy Policy* 75: 39–45.

Thøgersen, John, and Yanfeng Zhou. 2012. "Chinese Consumers' Adoption of a 'Green' Innovation: The Case of Organic Food." *Journal of Marketing Management* 28: 313–33.

Tilt, Bryan. 2007. "The Political Ecology of Pollution Enforcement in China: A Case from Sichuan's Rural Industrial Sector." *China Quarterly*, no. 192: 915–32.

Tilt, Bryan. 2013. "Industrial Pollution and Environmental Health in Rural China: Risk, Uncertainty and Individualization." *China Quarterly*, no. 214: 283–301.

Tilt, Bryan, and Qing Xiao. 2010. "Media Coverage of Environmental Pollution in the People's Republic of China: Responsibility, Cover-up and State Control." *Media, Culture & Society* 32: 225–45.

Tong, Jingrong. 2015. *Investigative Journalism, Environmental Problems and Modernization in China*. New York: Springer.

Uchida, Emi, Jintao Xu, and Scott Rozelle. 2005. "Grain for Green: Cost-Effectiveness and Sustainability of China's Conservation Set-Aside Program." *Land Economics* 81: 247–64.

Uchida, Emi, Jintao Xu, Zhigang Xu, and Scott Rozelle. 2007. "Are the Poor Benefiting from China's Land Conservation Program?" *Environment and Development Economics* 12: 593–620.

Uchida, Emi, Scott Rozelle, and Jintao Xu. 2009. "Conservation Payments, Liquidity Constraints and Off-Farm Labor: Impact of the Grain for Green Program on Rural Households in China." In *An Integrated Assessment of China's Ecological Restoration Programs*, ed. Runsheng Yin, pp. 131–57. Dordrect, Netherlands: Springer.

Unger, Jonathan, and Anita Chan. 1995. "China, Corporatism, and the East Asian Model." *Australian Journal of Chinese Affairs* 33: 29–53.

Van Rooij, Benjamin. 2003. "Organization and Procedure in Environmental Law Enforcement: Sichuan in Comparative Perspective." *China Information* 17: 36–64.

Van Rooij, Benjamin. 2006. "Implementation of Chinese Environmental Law: Regular Enforcement and Political Campaigns." *Development and Change* 37: 57–74.

Van Rooij, Benjamin. 2010. "The People vs. Pollution: Understanding Citizen Action against Pollution in China." *Journal of Contemporary China* 19: 55–77.

Van Rooij, Benjamin. 2012. "The Compensation Trap: The Limits of Community-Based Pollution Regulation in China." *Pace Environmental Law Review* 29: 701–45.

Van Rooij, Benjamin, and Carlos Wing-hung Lo. 2010. "A Fragile Convergence: Understanding Variation in the Enforcement of China's Industrial Pollution Law." *Law & Policy* 32: 14–37.

Van Rooij, Benjamin, Rachel Stern, and Kathinka Fürst. 2014. "The Authoritarian Logic of Regulatory Pluralism: Understanding China's New Environmental Actors." *Regulation & Governance* 10: 3–13.

Van Rooij, Benjamin, Qiaoqiao Zhu, Li Na, Wang Qiliang, and Zhang Xuehua. 2016. "Pollution Enforcement in China: Understanding National and Regional Variation." In *The Routledge Handbook of Environmental Policy in China*, ed. Eva Sternfeld, pp. 193–207. London: Routledge.

Wakeman, Frederic E. 1985. *The Great Enterprise: The Manchu Reconstruction of Imperial Order in Seventeenth-Century China, Volume 1*. Berkeley: University of California Press.

Wallace, Jeremy L. 2016. "Juking the Stats? Authoritarian Information Problems in China." *British Journal of Political Science* 46: 11–29.

Wang, Alex. 2006. "The Role of Law in Environmental Protection in China: Recent Developments." *Vermont Journal of Environmental Law* 8: 196–223.

Wang, Alex. 2013. "The Search for Sustainable Legitimacy: Environmental Law and Bureaucracy in China." *Harvard Environmental Law Review* 37: 365–440.

Wang, Ben Zhe, and Zhiming Cheng. 2015. "Environmental Perceptions, Happiness and Pro-Environmental Actions in China." *Social Indicators Research* 1: 1–19.

Wang, Chunmei, and Virginia Maclaren. 2012. "Evaluation of Economic and Social Impacts of the Sloping Land Conversion Program: A Case Study in Dunhua County, China." *Forest Policy and Economics* 14: 50–57.

Wang, Chunmei, Hua Ouyang, Virginia Maclaren, Y. Yin, B. Shao, A. Boland, and Y. Tian. 2007. "Evaluation of the Economic and Environmental Impact of Converting Cropland to Forest: A Case Study in Dunhua County, China." *Journal of Environmental Management* 85: 746–56.

Wang, Lei. 2016. "China's Crude Oil and Natural Gas Industry." Report for the Oil and Gas Conference, Denver, CO. Accessed December 29, 2016. http://www.theoiland gasconference.com/downloads_TOGC_2016/China-Oil-and-Gas-Lei-Wang-PhD .pdf.

Wang, Shaoguang. 2008. "Changing Models of China's Policy Agenda Setting." *Modern China* 34: 56–87.

Wang, Xinhong. 2014. "Open Environmental Information upon Disclosure Request in China: The Paradox of Legal Mobilization." Ph.D. diss., University of Turku.

Warwick, Mara, and Leonard Ortolano. 2007. "Benefits and Costs of Shanghai's Environmental Citizen Complaints System." *China Information* 21: 237–68.

Weber, Olaf. 2014. "Environmental, Social and Governance Reporting in China." *Business Strategy and the Environment* 23: 303–17.

Weller, Robert. 2008. "Responsive Authoritarianism." In *Political Change in China: Comparisons with Taiwan*, ed. Bruce Gilley, pp. 117–33. Boulder: Lynne Rienner.

Weller, Robert P. 2012. "Responsive Authoritarianism and Blind-Eye Governance in China." In *Socialism Vanquished, Socialism Challenged: Eastern Europe and China, 1989–2009*, eds. Nina Bandelj, and Dorothy J. Solinger, pp. 83–99. New York: Oxford University Press.

White, Gordon, Jude Howell, and Shang Xiaoyuan. 1996. *In Search of Civil Society: Market Reform and Social Change in Contemporary China*. Oxford: Clarendon Press.

Whiting, Susan. 2004. "The Cadre Evaluation System at the Grass Roots: The Paradox of Party Rule." In *Holding China Together: Diversity and National Integration in the Post-Deng Era*, eds. Barry J. Naughton and Dali L. Yang, pp. 101–19. New York: Cambridge University Press.

Wildau, Gabriel. 2015. "Smog Film Captivates Chinese Internet." *Financial Times*, March 2. Accessed December 29, 2016. https://www.ft.com/content/de190a92-cob0-11e4-876d-00144feab7de/.

Wong, Edward. 2015. "China Blocks Web Access to 'Under the Dome' Documentary on Pollution." *New York Times*, March 6. Accessed December 29, 2016. http://www.nytimes.com/2015/03/07/world/asia/china-blocks-web-access-to-documentary-on-nations-air-pollution.html?_r=0/.

Wong, Edward. 2016. "Clampdown in China Restricts 7,000 Foreign Organizations." *New York Times*, April 28. Accessed December 29, 2016. http://www.nytimes.com/2016/04/29/world/asia/china-foreign-ngo-law.html.

Wong, Stan Hok-wui, and Minggang Peng. 2015. "Petition and Repression in China's Authoritarian Regime: Evidence from a Natural Experiment." *Journal of East Asian Studies* 15: 27–67.

Worland, Justin. 2017. "It Didn't Take Long for China to Fill America's Shoes on Climate Change." *Time*, June 8. Accessed June 12, 2017. http://time.com/4810846/china-energy-climate-change-paris-agreement/.

World Bank. 2007. "Cost of Pollution in China: Economic Estimates of Physical Damages." Accessed December 21, 2016. http://edgar.jrc.ec.europa.eu/overview.php?v=CO2ts1990-2015.

Wu, Jing, Yongheng Deng, Jun Huang, Randall Morck, and Bernard Yeung. 2013. "Incentives and Outcomes: China's Environmental Policy." National Bureau of Economic Research (NBER) Working Paper. Washington, DC.

Xiao, Qiang. 2011. "The Battle for the Chinese Internet." *Journal of Democracy* 22: 47–61.

Xie, Lei. 2009. *Environmental Activism in China*. London: Routledge.

Xie, Lei. 2011. "China's Environmental Activism in the Age of Globalization." *Asian Politics & Policy* 3: 207–24.

Xie, Lei. 2015. "Political Participation and Environmental Movements in China." In *The International Handbook of Political Ecology*, ed. Raymond Bryant, 246–75. Northampton, MA: Edward Elgar.

Xie, Lei. 2016. "Environmental Governance and Public Participation in Rural China." *China Information* 30: 188–208.

Xinhuanet. 2016. "Jiceng huanbao jigou chuizhi guanli gaige da mu la kai 基层环保机构垂直管理改革大幕拉开 [Vertical Management Reform of Grassroots Environmental Protection Organizations]." Accessed December 29, 2016. http://news.xinhuanet.com/legal/2016-09/23/c_129294953.htm.

Xu, Zhigang, Jintao Xu, Xiangzheng Deng, Jikun Huang, Emi Uchida, and Scott Rozelle. 2006. "Grain for Green versus Grain: Conflict between Food Security and Conservation Set-aside in China." *World Development* 34: 130–48.

Xu, Jintao, Ran Tao, Zhigang Xu, and Michael T. Bennett. 2010. "China's Sloping Land Conversion Program: Does Expansion Equal Success?" *Land Economics* 86: 219–44.

Yang, Guobin. 2005. "Environmental NGOs and Institutional Dynamics in China." *China Quarterly*, no. 181: 46–66.

Yang, Guobin. 2009. *The Power of the Internet in China: Citizen Activism Online*. New York: Columbia University Press.

Yang, Guobin, and Craig Calhoun. 2007. "Media, Civil Society, and the Rise of a Green Public Sphere in China." *China Information* 21: 211–36.

Yang, Tseming, and Xuehua Zhang. 2012. "Public Participation in Environmental Enforcement with Chinese Characteristics? A Comparative Assessment of China's Environmental Complaint Mechanism." Vermont Law School Research Paper. South Royalton, VT.

Yao, Shunbo, Yajun Guo, and Xuexi Huo. 2010. "An Empirical Analysis of the Effects of China's Land Conversion Program on Farmers' Income Growth and Labor Transfer." *Environmental Management* 45: 502–12.

Yeh, Emily T. 2013. "The Politics of Conservation in Contemporary Rural China." *Journal of Peasant Studies* 40: 1165–88.

Yin, Shijiu, Linhai Wu, Lili Du, and Mo Chen. 2010. "Consumers' Purchase Intention of Organic Food in China." *Journal of the Science of Food and Agriculture* 90: 1361–67.

Yong, Ma. 2009. "Fahui shetuan zuzhi zuoyong, tuidong huanjing gongyi susong 发挥社团组织作用，推动环境公益诉讼 [The Role Social Organizations Can Play in Pushing Forward Environmental Public Interest Litigation]." In *Zhongguo huanjing fazhi* 中国环境法制 [Environmental Law in China], ed. Jiwen Chang pages?. Beijing: Falu chubanshe.

Yu, Xiang, and Alex Y. Lo. 2015. "Carbon Finance and the Carbon Market in China." *Nature Climate Change* 5: 15–16.

Yu, Xueying. 2014. "Is Environment 'A City Thing' in China? Rural–Urban Differences in Environmental Attitudes." *Journal of Environmental Psychology* 38: 39–48.

Zeng, S.X., X.D. Xu, Z.Y. Dong, and Vivian W.Y. Tam. 2010. "Towards Corporate Environmental Information Disclosure: An Empirical Study in China." *Journal of Cleaner Production* 18: 1142–48.

Zhan, Xueyong, and Shui-Yan Tang. 2013. "Political Opportunities, Resource Constraints and Policy Advocacy of Environmental NGOs in China." *Public Administration* 91: 381–99.

Zhang, Bing, Hanxun Fei, Pan He, Yuan Xu, Zhanfeng Dong, and Oran R. Young. 2016. "The Indecisive Role of the Market in China's SO_2 and COD Emissions Trading." *Environmental Politics* 25: 875–98.

Zhang, Bing, Yan Yang, and Jun Bi. 2011. "Tracking the Implementation of Green Credit Policy in China: Top-Down Perspective and Bottom-Up Reform." *Journal of Environmental Management* 92: 1321–27.

Zhang, Bing, Hui Zhang, Beibei Liu, and Jun Bi. 2013. "Policy Interactions and Underperforming Emission Trading Markets in China." *Environmental Science & Technology* 47: 7077–7084.

Zhang, Da, Valerie J. Karplus, Cyril Cassisa, and Xiliang Zhang. 2014. "Emissions Trading in China: Progress and Prospects." *Energy Policy* 75: 9–16.

Zhang, Joy Y., and Michael Barr. 2013. *Green Politics in China: Environmental Governance and State-Society Relations.* London: Pluto Press.

Zhang, Lei, Guizhen He, and Arthur P.J. Mol. 2015. "China's New Environmental Protection Law: A Game Changer?" *Environmental Development* 13: 1–3.

Zhang, Lei, Arthur P.J. Mol, and Guizhen He. 2016. "Transparency and Information Disclosure in China's Environmental Governance." *Current Opinion in Environmental Sustainability* 18: 17–24.

Zhang, Lei, Arthur Mol, Guizhen He, and Yonglong Lu. 2010. "An Implementation Assessment of China's Environmental Information Disclosure Decree." *Journal of Environmental Sciences* 22: 1649–1656.

Zhang, Man, Qiao Hui, Wang Xu, Ming-zhe Pu, Zhi-jun Yu, and Feng-tian Zheng. 2015. "The Third-Party Regulation on Food Safety in China: A Review." *Journal of Integrative Agriculture* 14: 2176–2188.

Zhang, Xin, Xiaobo Zhang, and Xi Chen. 2015. "Happiness in the Air: How Does a Dirty Sky Affect Subjective Well-Being?" Institute for the Study of Labor (IZA) discussion paper. Bonn, Germany.

Zhang, Xuehua. 2008. "Enforcing Environmental Regulations in Hubei Province, China: Agencies, Courts, Citizens." Ph.D. diss., Stanford University.

Zhang, Xuehua. 2010. "Green Bounty Hunters: Engaging Chinese Citizens in Local Environmental Enforcement." *China Environment Series* 11: 137–153.

Zhang, Xuehua. 2014. "Judicial Enforcement Deputies: Causes and Effects of Chinese Judges Enforcing Environmental Administrative Decisions." *Regulation & Governance* 10: 29–43.

Zhang, Xuehua, Leonard Ortolano, and Zhongmei Lü. 2010. "Agency Empowerment through the Administrative Litigation Law: Court Enforcement of Pollution Levies in Hubei Province." *China Quarterly*, no. 202: 307–326.

Zhang, Yanshuang. 2016. "Digital Environmentalism: A Case Study of $PM_{2.5}$ Pollution Issue in Chinese Social Media." Working Paper, University of Queensland.

Zhang, Zhiming, John Aloysius Zinda, and Wenqing Li. 2017. "Forest Transitions in Chinese Villages: Explaining Community-Level Variation under the Returning Forest to Farmland Program." *Land Use Policy* 64: 245–257.

Zhang, Zhongxiang. 2015. "Carbon Emissions Trading in China: The Evolution from Pilots to a Nationwide Scheme." *Climate Policy* 15: S104–S126.

Zhao Ping, Min Dai, Chen Wanqing, and Li Ni. 2010. "Cancer Trends in China." *Japanese Journal of Clinical Oncology* 40: 281–285.

Zhao Xiaofan and Leonard Ortolano. 2010. "Implementing China's National Energy Conservation Policies at State-Owned Electric Power Generation Plants." *Energy Policy* 38: 6293–6306.

Zhong, Yang, and Wonjae Hwang. 2015. "Pollution, Institutions and Street Protests in Urban China." *Journal of Contemporary China* 25: 216–232.

Zhou, Decheng, Shuqing Zhao, and Chao Zhu. 2012. "The Grain for Green Project Induced Land Cover Change in the Loess Plateau: A Case Study with Ansai County, Shanxi Province, China." *Ecological Indicators* 23: 88–94.

Zhou, Li-an. 2007. "Governing China's Local Officials: An Analysis of Promotion Tournament Model." *Economic Research Journal* 7: 36–50.

Zhou, Xueguang. 2010. "The Institutional Logic of Collusion among Local Governments in China." *Modern China* 36: 47–78.

Zhou, Xueguang, Hong Lian, Leonard Ortolano, and Yinyu Ye. 2013. "A Behavioral Model of "Muddling through" in the Chinese Bureaucracy: The Case of Environmental Protection." *China Journal* 70: 120–147.

Zinda, John Aloysius, Christine J. Trac, Deli Zhai, and Stevan Harrell. 2017. "Dual-function Forests in the Returning Farmland to Forest Program and the Flexibility of Environmental Policy in China." *Geoforum* 78: 119–132.

Printed in the United States
By Bookmasters